Lyndon B. Johnson School of Public Affairs
Policy Research Project Report
Number 67

Pesticides and Worker Health in Texas

A report by the
Policy Research Project on Pesticide Regulation in Texas
The University of Texas at Austin
1984

Library of Congress Card Number: 84-82100
ISBN: 0-89940-669-6

Foreword

The Lyndon B. Johnson School of Public Affairs has established inter-disciplinary research on policy problems as the core of its educational program. A major part of this program is the nine-month policy research project, in the course of which two or three faculty members from different disciplines direct the research of ten to twenty graduate students of diverse backgrounds on a policy issue of concern to a government agency. This "client orientation" brings the students face to face with administrators, legislators, and other officials active in the policy process, and demonstrates that research in a policy environment demands special talents. It also illuminates the occasional difficulties of relating research findings to the world of political realities.

This analysis of pesticides and worker health in Texas is the product of a policy research project conducted at the LBJ School in academic year 1983-84. Publication was funded and overseen by the Texas Department of Agriculture, which commissioned the study. A second volume on regulating pesticides is being published in the same manner as a companion volume. Additional funds for publication were provided by the Lyndon Baines Johnson Foundation.

The curriculum of the LBJ School is intended not only to develop effective public servants but also to produce research that will enlighten and inform those already engaged in the policy process. The project that resulted in this report has helped to accomplish the first task; it is our hope and expectation that the report itself will contribute to the second.

Finally, it should be noted that neither the LBJ School nor The University of Texas at Austin necessarily endorses the views or findings of this study.

Max Sherman
Dean

PROJECT PARTICIPANTS

Participating Students

Cyral Miller   Virginia Raymond
Anne Dee Rader   Gary Watts

Project Directors

Susan G. Hadden, LBJ School of Public Affairs
Thomas O. McGarity, University of Texas School of Law

Table of Contents

PESTICIDES AND WORKER HEALTH IN TEXAS

1. INTRODUCTION

> Within the agricultural industry it is the farm laborer who
> experiences most of the occupational disease from pesticides . . . .
> Because of inadequate education, language problems, migratory status,
> substandard health, and poor hygiene, these workers are the least
> likely of any group to be able to work safely with farm chemicals.[1]

Although this statement was made over twenty years ago, it still holds true.

The purpose of this report is to examine critically the data about worker

health and pesticides and to identify policy options that are available for

treating those problems that are identified. This is a complex subject with a

broad literature, but one that has astonishingly not been gathered into one

place. Another purpose of this report, therefore, is to provide an

introduction to the existing literature on pesticide exposure and health.

Pesticides are used quite extensively in both agricultural and urban

settings in the United States. In 1980, Americans used almost 1.2 billion

pounds of pesticides.[2] The United States Department of Agriculture (USDA)

estimates that 225 to 250 million acres, or about two-thirds of American crop

lands are treated annually with pesticides.[3] Moreover, of the $5 billion to $6

billion per year that U.S. farmers spend on pest control, the majority goes

toward purchasing pesticides.[4] USDA estimates that, excluding the cost of

land, pesticides account for 2 to 13 percent of the total production costs for

major field crops.[5]

While the agricultural sector uses the majority of the pesticides sold in

the United States, urban areas also seem to use pesticides on a broad scale.

For example, a 1979 study by the Environmental Protection Agency (EPA) found

that 90.7 percent of the households surveyed used pesticides in the house, garden, or yard.[6] Urban dwellers not only use pesticides commonly, but they appear to use large quantities of them. The authors of a 1972 study on pesticide use in three urban areas (Philadelphia, Pennsylvania; Dallas, Texas; and Lansing, Michigan) estimated the average deposit of active pesticide ingredients in those areas to be between 5.3 and 10.6 pounds per acre.[7]

According to a 1974 EPA study, Texas is ranked second in the nation in its use of pesticides. However, since Texas does not have a system to collect data on the quantity of pesticides used, it is nearly impossible to determine the exact amount used in the state. Estimates of pesticide use in Texas range widely. The 1974 EPA study estimated that at least 89 million pounds of pesticides were applied in Texas,[8] while a more recent study estimated that in 1977 between 100 and 150 million pounds of pesticides were used.[9] Pesticide use in Texas appears to be on an upward trend.[10]

Most of the pesticides sold in Texas are used in the agricultural sector. Texas agriculture is ranked second in the nation in the total cash receipts for farm production, and first in the nation in total number of farms and amount of farm and ranch acreage.[11] A survey conducted in 1980 indicated that 54 percent of Texas farm and ranch acreage received pesticide applications.[12]

Such extensive use of an expensive agricultural input would not occur unless pesticides offered great benefits. Indeed, they have proven to be a great boon by reducing crop damage caused by various pests, as well as by killing pests that transmit diseases. However, many pesticides have also placed a great burden on people and the environment. Some pesticides have been shown to have acute and chronic effects on humans, including poisonings at low doses, cancer, genetic mutations, and birth defects. Similarly, some pesticides have been found to damage aquatic and terrestrial ecosystems. In order to protect the public and the environment, yet still benefit from the use of pesticides, federal and state governments have acted to regulate their use.

The lead agency responsible for regulating the use of pesticides in Texas is the Texas Department of Agriculture (TDA). In autumn 1983, TDA asked a study team at the Lyndon B. Johnson School of Public Affairs to assess a variety of policies that the State can use to regulate pesticides and prepare a report outlining the advantages and disadvantages of each policy option. In the course of that work, one subgroup found that there was an especially large body of widely scattered literature and an especially broad range of opinion on the effects of pesticides on the health of agricultural workers, the group most frequently and heavily exposed to them. This group believed that it would assist both TDA and others interested in this question if a work were made available that provided more technical detail than the general report and pulled together the scattered studies. This volume therefore contains an overview of what is known about the health effects of pesticides and also assesses programs now in place to monitor and reduce danger to health. The companion volume, *Regulating Pesticides in Texas*, contains a shorter version of this report and assessments of other issues in pesticide regulation in Texas, including licensing, urban use, pesticide residues in food, and enforcement. That volume is also available from the Lyndon B. Johnson School of Public Affairs at The University of Texas at Austin or from the Texas Department of Agriculture.

1.1 Overview of the Report

The most important theme of this report is that policy can be made even though the information that is available is open to a variety of interpretations. Establishing a clear cause-and-effect relationship between pesticide exposure and health effects requires carefully controlled scientific studies which cannot be conducted under normal field or working conditions. Since there is no requirement that pesticide poisonings be reported, since the symptoms of pesticide poisonings imitate those of many other diseases, and since the health of agricultural workers depends on many factors other than pesticide exposure, obtaining sound data is extremely difficult. Therefore, there are many unanswered scientific questions that underlie policy. Nevertheless, many people believe that it is important to take action on the basis of the evidence we now have, which suggests that there are serious health effects from at least some pesticide exposures. This conclusion is not based on any one study, because almost all of them have at least some methodological problems; however, taken all together, the sheer mass of the data indicates that pesticides are implicated in a variety of health problems, ranging from short-term rashes to cancer and birth defects.

Since this report was written for the Texas Department of Agriculture, it focuses on fieldworkers. Data from studies of workers in pesticide manufacturing plants are not included. Although these studies often show that pesticides cause health problems, critics argue that they are not relevant to worker exposure in the fields, where the pesticide is much more dilute and where workers are outside rather than in an enclosed factory. As we shall see, data from field exposure suggest that fieldworkers also experience serious health effects.

The report is organized as follows: Chapter 2 classifies pesticides in various ways and relates those classifications to health effects. The bulk of the chapter considers the evidence for acute and chronic effects of pesticide exposure, and also presents some more qualitative evidence concerning exposure by different farmworker groups. Chapter 3 presents information specifically about Texas. It first describes farmworkers in Texas and health care available to them. Then it presents a variety of data about workers' exposures to pesticides, including both quantitative data and anecdotal evidence. The final section of the chapter offers a correlational analysis of some aggregate data available in Texas. Chapter 4 turns to the policy implications of the data presented in chapters 2 and 3. It considers current policies, presents some policy options, and discusses some new initiatives at both the federal and state levels.

2. HEALTH EFFECTS OF PESTICIDE EXPOSURE

2.1 How Pesticides Affect Health

The potential health hazard of a pesticide depends on five factors:

1. the inherent toxicity of the active ingredient;

2. the chemical and physical properties of the active ingredient;

3. the duration of exposure;

4. the dose and/or concentration;

5. the route of absorption.[13]

By design, pesticides inhibit or interfere with certain basic physiologic functions of living organisms. The inherent toxicity to humans is a function of the relative physical similarities between people and pests, combined with the fact that most pesticides are developed to kill a broad spectrum of organisms.

Pesticides can be categorized according to chemical structure, which determines the nature of their toxic effects. Table 1 gives some examples of pesticides included in three of the most important pesticide families -- organophosphates, chlorinated hydrocarbons, and carbamates -- and two older families.

Organophosphates poison insects and mammals by the phosphorylation of the enzyme acetylcholinesterase at nerve endings. In animals and humans, acetylcholine is the transmitter chemical for impulses in the central nervous system. Normal body functioning requires the rapid degradation of acetylcholine by the enzyme cholinesterase. If acetylcholine cannot be

Table 1. Examples of Four Pesticide Types

Arsenical compounds
        Arsenic acid   Disodium methyl arsonate
        Arsenic trioxide   Lead arsenate
        Cacodylic acid   Methane arsonic acid
        Calcium acid   Monoammonium methyl
          methanearsonate   arsonate
        Calcium arsenate   Monosodium methyl
        Calcium arsenite   arsonate
        Copper acetoarsenite   Sodium arsenate
        Copper arsenite   Sodium arsenite

Chlorinated hydrocarbons
        Benzene hexachloride (BHC)   Kepone
        Chordane   Heptachlor
        DDT   Hexachlorobenzene (HCB)
        Dicofol (Kelthane)   Lindane (Isomer of BHC)
        Dienochlor   Mirex
        Dieldrin   Thiodan
        Endrin   Toxaphene

Organophosphates
        Abate   Ethion
        DDVP   Fenthion (Baytex)
        Diazanon   Gardona
        Dicathon   Malathion
        Dimethoate   Naled (Dibrom)
        Dursban   Parathion
        EPN

Carbamates
        Baygon   Vapam
        Carbaryl (Sevin)   Zectran
        Thiram

Pyrethrins
        Allethrin   Fenpropanate
        Barthrin   Fenvalerate
        Bioresmethrin   Permethrin
        Cypermethrin   Phthalthrin
        Decamethrin   Resmethrin
        Fenothrin

Sources: U.S. Environmental Protection Agency, "Recognition and
      Management of Pesticides Poisonings," Technical Report
      EPA 540/9/80/005, January 1982; National Institute
      for Occupational Safety and Health,"Occupational Exposure
      during the Manufacture and Formulation of Pesticides,"
      July 1978.

degraded for some reason, there is no cessation of stimulation. Sustained nervous stimulation can be fatal to the organism, as normal control over bodily functions is lost.[14]   Pesticides that work by preventing the cholinesterase from breaking down acetylcholine are called "cholinesterase-inhibiting."

Chlorinated hydrocarbons interfere with the transmission of nerve impulses, disrupting the normal function of the nervous system, especially the brain. Most of these chemicals are absorbed efficiently from the gut and the skin. Chlorinated hydrocarbons are known for their persistence in the environment; they can remain active for two to twenty years after application. These chemicals are hydrophobic; they are not water soluble and are readily stored in adipose (body fat) tissue. This ability to be stored in living tissue promotes bioaccumulation, which is the tendency for chlorinated hydrocarbons to become more concentrated in tissue as it moves upward through the food chain. Because of their chemistry, organochlorines are especially prone to be excreted in the milk of lactating women, and may remain in the body for years.

Carbamates act upon acetylcholinesterase, inhibiting this enzyme's normal function of breaking down acetylcholine. As with organophosphates, this leads to an accumulation of acetylcholine. The carbamyl-enzyme complex, however, breaks down readily.   Thus the inhibiting effect of carbamates is "reversible."   This makes the confirmation of poisoning by plasma or cholinesterase testing very difficult.   Following the absorption of extraordinary amounts of these chemicals, enzyme levels typically return to normal within a few minutes or hours. In this sense carbamates are less toxic than organophosphates. The prognosis for recovery is generally better.   Like the organophosphates, carbamates can be absorbed via inhalation, ingestion, and dermal contact.[15] These general descriptions may obscure the fact that a difference of just one bond between two substances in the same chemical family may mean that they have completely different characteristic toxic effects. It is also important to realize that pesticides with high acute toxicity do not necessarily pose chronic hazards, and vice versa. Rather than discussing health effects further according to these classes or according to the five factors, therefore, we present them according to the intensity of exposure and the immediacy with which symptoms are manifested.

2.2 Effects of Pesticide Exposure

The possible effects of pesticide exposure can be divided into three categories:

1. acute exposure, or immediate identifiable response, such as poisoning or topical injuries;

2. chronic low exposure, such as occurs from exposure to pesticide residues on food, which may cause long-delayed health effects;

3. chronic high exposure, the long-term effects of more intense exposure to pesticides, which often occurs occupationally.

Occupationally exposed workers, including pesticide manufacturers and agricultural workers, are at risk in all three categories. The general public is also at risk in the first two categories. Pesticide accidents in the home or garden pose a risk of acute poisoning to individual urban users, while the public in general is chronically exposed to small amounts of pesticides in water, air, food, and clothing. As noted, this report focuses on the acute and high chronic risks posed to agricultural workers, because their exposures are more likely and more intensive. In some sense, the experiences of agricultural workers serve as an indicator or possible long-term effects of low-level exposure among the general population.

2.2.1 Acute Poisoning

The symptoms of acute systemic poisoning are well understood and well known. Accurate diagnosis, however, is difficult, because symptoms of pesticide exposure mimic those of many other diseases, including cerebrovascular disease, cardiovascular disease, heat stroke, and pneumonia.

Anticholinesterase pesticide poisonings can be grouped into four general classes, according to severity. These are:

1. *Latent poisoning*. There are no clinical manifestations; diagnosis depends on the estimation of serum cholinesterase activity, which is depressed to 50 to 90 percent of normal values. Treatment is unnecessary, but observation for at least six hours is advisable, since clinical manifestations of poisoning may progress.

2. *Mild poisoning* causes symptoms such as headache, fatigue, dizziness, blurred vision, excessive sweating and salivation, and abdominal cramps or diarrhea. (These symptoms are also shared by many illnesses unrelated to pesticides, such as influenza, heat stroke, and gastroenteritis.) In a mild pesticide poisoning, serum cholinesterase activity is 20 to 50 percent of normal value.

3. *Moderate poisoning* can cause all of the above symptoms, but in addition the patient cannot walk, has chest discomfort and difficulty talking, and exhibits marked miosis and muscle twitching. (These symptoms might reasonably be mistaken for such conditions as pneumonia, myocardial infarction, and encephalitis.)[16] Serum cholinesterase activity is only 10 to 20 percent of normal value.

4. *Severe poisoning* results in unconsciousness, local or generalized seizures, respiratory difficulty, and a cholinergic crisis in which serum cholinesterase activity is lower than 10 percent of normal value. (In these cases, several alternative causes of coma enter into the differential diagnosis.)

In addition to these systemic symptoms, topical (local) effects of pesticide exposure, such as dermatitis or eye problems, are quite common among agricultural workers. Dermatitis (inflammation of the skin) from pesticides can result from exposure to primary irritants, which can cause a chemical burn or severe irritation on almost anyone's skin, and from contact sensitizers, which affect only a few individuals who have become "sensitized," or allergic, to the material. Some areas of the body are more susceptible than others. The genitalia and eyelids are particularly vulnerable. Dermatitis from a contact sensitizer may be difficult to diagnose as pesticide-related, because the reaction may occur up to a week after exposure.[17] Contamination of apparently unexposed parts of the body by materials on the hands may also confound diagnosis.

Pesticide-related eye problems are also common among agricultural workers. Eye injuries can result from accidentally splashing or spilling the material into the eye, exposure to pesticide drift, and rubbing the eyes with contaminated hands. Such injuries are particularly likely in pesticide mixers, loaders, and applicators because of their risk of exposure to the pesticide concentrate. Additionally, as already noted, anticholinesterase pesticides have a miotic effect on the eyes.

The symptoms associated with acute poisoning are widespread. In a 1980 survey of 469 farmworkers in South Florida, for example, about 75 percent indicated that they had experienced one or more anticholinesterase poisonings symptoms (skin rash, headache, dizziness, excessive sweating, etc.).[18] About 38 percent had experienced moderate poisoning symptoms, including chest pains and inability to walk; about 24 percent had experienced severe poisoning symptoms such as bronchial secretions and convulsions. (Definitions of "severe" vary. These authors included drowsiness under severe effects.) Almost 45 percent of those with one or more symptoms were sufficiently concerned to consult a physician, and about half of these farmworkers received a pesticide poisoning diagnosis.

Proportionately more women than men complained of such symptoms as excessive sweating, drowsiness, swelling, blurred vision, nausea and vomiting, and headaches. If this is because more women than men are being made ill by their exposure to pesticides, it may indicate that women in the field are more susceptible than men because tolerances and reentry times are set for the greater body weights of men or because women may be wearing different types of clothing than men. An alternative explanation is that the women in the survey were simply more willing to report their physical complaints, because they may, in general, have more contact with the health care delivery system through maternal and child health care providers.

As one would expect, symptoms of acute poisoning are usually associated with direct exposure in the field. A majority (75 percent) of the farmworkers who experienced one or more symptoms of pesticide poisoning were working in or near an area that was being sprayed at the time they became ill. Most (90 percent) were aware of inhaling pesticide vapors, and many (75 percent) were also aware of pesticide drift or residue settling on their clothing at the time they became ill.

Verifying pesticide exposure and relating even obvious acute symptoms to particular pesticides is difficult for a variety of reasons. First, acute toxicity varies with a number of genetic and environmental factors, including route of exposure (e.g., dermal vs. oral), age, sex, and genetic susceptibility, ambient temperature, diet, absence or presence of other active compounds, and duration of exposure.[19] Second, as noted, symptoms can mimic other disorders such as influenza. The tests used to diagnose the presence of pesticide poisoning are also problematic; this is discussed below. Third, farmworkers' generally low standard of health reduces their concerns about the specific effects of pesticide exposure.[20] Finally, workers who are recruited by crew leaders and driven out to the field may well not be able to identify the fields in which they worked; this hampers attempts to confirm exposures to specific pesticides after the fact.[21]

Because of the lack of a reliable reporting system and the fact that many poisonings go unrecognized, there are no adequate estimates (or even ranges) of the current number of poisonings per year. In 1970, the U.S. Department of Health, Education and Welfare estimated that 800 persons were killed and 800,000 injured annually as a result of pesticides.[22] Data from various states also suggest in sum a very high poisoning rate. Dr. Ephraim Kahn believes that California's monitoring system records only 1 to 2 percent of all the poisonings in that state.[23]

A 1971-76 study of hospital-admitted pesticide poisonings represents the only systematic nationwide study to date.[24] In 1971, the estimated number of such poisonings was 2,596; in 1976, the estimated number of hospital cases was 3,010. The use of hospital-based data (standardized diagnosis codes related to pesticides) provides for the greatest uniformity and accuracy in identifying and studying cases, but seriously underestimates the total number of acute pesticide poisonings in the country. That is because the number of poisonings treated outside hospitals and the number not treated at all are simply unknown. The extent of underestimation is also unknown. Studies conducted in South Carolina indicate that for every hospital-admitted case in that state a physician treats fifteen office cases.[25]

2.2.1.1 Hypersensitivity

One special category of effects of pesticide exposure is hypersensitivity.
This condition makes people unusually susceptible to acute effects of
pesticide exposure; it is not known whether it makes them more susceptible to
delayed effects. In general, long-term exposure to pesticides tends to make
peple sensitive to the chemical. A farmworker who has been exposed to
pesticides for twenty years might react to a relatively small dose of
pesticide that would not pose a similar problem to a person who lacked the
history of sustained exposure.[26] Sensitization can result from skin contact
or inhalation. While sensitization usually occurs after extended exposure to
relatively high concentrations of pesticides, allergic responses can result
following "very low" concentration exposure.[27]

According to NIOSH, a number of pesticides have been known to cause skin
hypersensitization in humans. Postincident sensitization has been documented,
for instance, in cases involving naled and allidochlor. In one set of
experiments, it was discovered that a single exposure to a "nonirritating 1
percent solution of malathion" induced reactions of "great" intensity in
almost one-half of those subjects who had been previously exposed to a 10
percent solution of malathion.[28]

2.2.2 Chronic Effects

It is frequently asserted that the effects of long-term contact with
pesticides are unknown or difficult to determine. Most studies in the field
and all efforts toward reporting and estimating cases have indeed focused on
documentation of acute pesticide poisonings. However, a substantial body of

human research has been conducted during the last decade which suggests links between chronic pesticide exposure and serious delayed effects, which are usually termed "chronic." Unlike most acute toxic effects, these chronic effects tend to be irreversible. The liver is the organ most frequently damaged; however irreversible damage to the central nervous system, the peripheral nerves, the kidneys, and other organs may occur. In addition to neurotoxicity, the major chronic effects may be classified as carcinogenetic, mutagenetic, teratogenetic, and reproductive.

2.2.2.1 Carcinogenicity

By 1978, NIOSH had reviewed 113 pesticides to assess their carcinogenic effect on humans and animals. Twenty-six (23 percent) of those reviewed were categorized as suspected occupational carcinogens. These pesticides are detailed in table 2. "Suspected" means that laboratory studies found a statistically significant relationship between pesticide exposure and tumor development in one or more mammalian species. Carcinogenic data were considered inconclusive for another 27 (24 percent) of the pesticides reviewed. The remaining 60 pesticides (53 percent) evidenced no carcinogenic effect. About 1300 pesticides have not been reviewed.[29]

Table 2. Partial List of EPA Restricted-Use Pesticides Due to Concerns for Human Health Effects

Pesticide	Acute Toxicity Dermal	Inhalation	Oral	Hazard Dermal	Inhalation	Accident History
Acrolein (Aqualin)					X	
Acrylonitrile (Acrylon)						X
Aldicarb (Temik)						X
Allyl alcohol	X					
Aluminum phosphide						

18

(Phostoxin)					X	
Azinphosmethyl (Guthion)					X	
Calcium Cyanide (Cyanogas)					X	
Carbofuran (Furadan)		X				
Chlorfenvinphos (Supona)	X					
Chlorophacinone (Rozol)			X			
Clonitralid (Bayluscide)		X				
Demeton (Systox)	X		X			
Dioxathion (Delnav)	X					
Disulfoton (Di-Syston)	X	X				
Endrin	X					
EPN	X		X			
Ethoprop (Mocap)	X					
Ethyl parathion	X				X	
Fenamiphos (Hemacur)	X					
Fensulfothion (Dasanit)	X	X				
Fluroacetamide (1081)			X			
Fonofos (Dyfonate)	X					
Hydrocyanic acid (HCN)					X	
Magnesium phosphide					X	
Methamidophos (Monitor)	X					
Methomyl (Lannate, Nudrin)						X
Methyl bromide						X
Methyl parathion	X					X
Mevinphos (Phosdrin)	X					
Monocrotophos (Azodrin)	X					
Paraquat						X
Phorate (Thimet)	X					
Phosphamidon (Dimecron)	X				X	
Sodium cyanide (Cymag)					X	
Sodium fluoroacetate (1080)			X			X
Strychnine			X			X
Sulfotep (Bladafum)					X	
Tepp				X	X	
Zinc phosphide			X			

Source: George W. Ware, *Pesticides: Theory and Application* (San Francisco: W. H. Freeman, 1983), pp. 190-91.

Field studies also link chronic exposure to pesticides with various types of
cancer.   A study of six types of cancer mortality in Iowa from 1971 to 1978,
found that farmers had "significantly elevated" mortality rates compared to
nonfarmers.[30] Further studies examining mortality rates in conjunction with
selective farming practices (pesticide use, milk products sold,   etc.)
indicate that elevated rates of multiple myeloma and non-Hodgkin's lymphoma
are associated with insecticide and herbicide use.[31] Evidence from Nebraska
and Iowa also suggests that leukemia mortality in farmers is elevated only in
relatively recent birth cohorts, suggesting that "modern farming methods . . .
merit further investigation. This concern is reinforced by evidence that
herbicide use in Iowa and insecticide use in Nebraska are not associated with
leukemia mortality in farmers born before 1890."[32]

Other studies associate pesticide exposure with lung cancer.   A study of
workers involved with the manufacture and packaging of pesticides (employed
during some period between 1946 and 1977) found a significantly increased
mortality ratio for lung cancer and anemia in males. A dose-response effect
was suggested for lung cancer mortality, which increased with length of
arsenical exposure (but not for nonarsenicals).[33] A West German study of
pesticide-exposed male agricultural workers found that the mortality ratio for
lung cancer was significantly higher for that group than for the general male
population.   It is noteworthy that the smoking habits of the exposed workers
did not differ from those of the general population. Furthermore, a positive
correlation between duration of employment and mortality due to lung cancer
suggests a dose-effect relation.[34]

In 1982, the California Department of Health Services reported the results of an epidemiological study that compared DBCP drinking water contamination and mortality from specific cancers (esophageal, stomach, liver, kidney, female breast, and lymphoid leukemia) in Fresno County, California, from 1970 to 1979. DBCP (1,2-dibromo-3-chloropropane) is a nematocide which until its suspension by the State in 1977 was used widely on grapes and tree fruit in California. DBCP had been linked to sterility and lowered sperm counts as well as cancers in animal studies. With the discovery that DBCP had permeated groundwater in many areas, it was banned from the continental United States in 1978.

The report claims to be the first effort to examine the possible association between cancer and an environmental exposure to DBCP. The study found statistically significant relationships between an increasing exposure to DBCP and male stomach cancer, total stomach cancer, and total lymphoid leukemia. However, the study is careful to note that the long latency period of the diseases, incomplete data, and a lack of controls for other variables preclude implicating DBCP as a cause of cancer based solely on the mortality data analyzed. Therefore, further investigation and evaluation are suggested.[35]

Other studies have not found a relationship between pesticide exposure and cancer. One, for example, conducted a proportionate mortality analysis of all deaths recorded among California resident farmworkers and farm owners and managers from 1978 to 1979. Because this study was limited to people who claimed "farmworker" as their principal occupation, it probably excluded the majority of migrant and non-resident farmworkers. The study found that farmworkers suffer more than others from accidents and certain other diseases.

However, while some cancer mortality rates were slightly elevated for some groups, there was no significant "evidence of excess mortality due to any of the malignant causes of death" for farmworkers or farm owners and managers. The report adds a clarifying note: "Failure in this study to detect many significant excesses of deaths due to specific cancers and other diseases (which may indeed be related to occupational exposure to pesticides) must be evaluated in consideration of the fact that a real association could be diluted by not studying only deaths among 'high risk' agrigultural workers, e.g., among migrant workers."[36] Two other studies also found that cancer mortality among "occupationally exposed" workers (including agricultural workers, fumigators, etc.), although elevated, was not significantly different from that of nonexposed groups.[37]

2.2.2.2 Mutagenicity and Teratogenicity

Mutagens cause alteration of DNA and teratogens cause malformation of the fetus. These changes are even more difficult to detect and to assign to particular pesticide exposures than carcinogenic effects. For example, a number of pesticides, including methyl parathion and captan, have caused genetic mutations under laboratory conditions. However there is no evidence that any pesticide has actually caused mutagenesis in humans.[38]

A large amount of anecdotal evidence associates pesticide exposure with birth defects. Empirical evidence comes from a hospital chart-review study of an agricultural community in California.[39] In cases where one or both parents were agricultural workers, offspring showed a higher proportion of limb malformations -- 5.2 defects per 1,000 live births -- than other offspring in the community -- 1.3 defects. Both rates are above the value of 0.40, an expected value based on the western United States. It should be noted that underestimation of defects could occur, because defects may be manifested after the infant has left the hospital. The study is also limited by its inability to control for the many other factors contributing to the workers' health.[40] Another study is being initiated in California. The study will compare birth outcomes among agricultural worker families with outcomes among a sample of nonagricultural women "matched" in terms of age, ethnicity, nutrition and other factors to the extent feasible.[41]

There is evidence which suggests that eight pesticides (thiram, aldrin, dieldrin, endrin, folpet, captan, captafol, and 2,4,5-T) pose a potential teratogenic problem in humans. In addition, there is somewhat weaker evidence that four other pesticides (parathion, dichlorvos, diazanon, and phosmet) may also pose a birth defect problem. NIOSH recommends that all twelve of these pesticides should be avoided by women of childbearing age.[42]

2.2.2.3 Reproductive Effects

There are many studies of long-term effects on reproduction. A 1981 report for the President's Council on Environmental Quality summarizes nine studies, and notes increased frequencies of impotence, chromosome aberrations, infertility, miscarriage, toxemia, and other adverse effects on reproduction among both men and women occupationally exposed to various pesticides.[43] Another study, of workers in a pesticide formulation plant, concluded that "[t]he relationship between duration of chemical exposure and sperm count was striking." The study, initiated because of increasing awareness of infertility among workers in a special agricultural chemical division, found evidence of severely decreased sperm counts among exposed workers relative to nonexposed men.[44] Kepone, mirex, aldrin, DDT, dieldrin, 2,4,5-T, carbaryl, heptachlor, and crufomate have been shown to cause decreased fertility in animals.[45]

2.2.2.4 Other Chronic Disorders

Chronic effects of pesticide exposure have also been associated with several other illnesses.   The results of a Danish study of fruitgrowers and farmers during the spraying season indicate that pesticides "can give rise to a lung disease, 'biocide [pesticide] lung,' which comprises (1) pneumonia, . . . and (2) chronic progressive lung fibrosis."[46] Another study cites links between organophosphates and carbamates and neurological disorders, behavioral effects, and decreased resistance to viruses and respiratory ailments.[47] Finally, a fourteen-state study of disease incidence rates among occupationally exposed workers and nonexposed controls finds "apparent associations   between   [organochlorine pesticide levels] . . . and the subsequent appearance of hypertension, arteriosclerotic cardiovascular disease, and possibly diabetes."[48]

2.3 HAZARDS TO SPECIAL GROUPS

Pesticide poisonings are most likely to occur in three main groups of people: applicators, "pickers," and children or other members of an exposed worker's family.   The hazard to applicators results from the dilution and application of the pesticide concentrate. Hazards exist, therefore, in both mixing and applying the pesticide. A 1974 analysis of physicians' reports in California indicated that most of the pesticide poisonings occurred in individuals who mixed, loaded, or applied the chemicals.   Fieldworkers experienced fewer cases of (identifiable) pesticide illnesses.[49]   These hazards are so obvious that we do not consider them further here.

Once a pesticide has been diluted to its final concentration and is applied to the crop, the pesticide residue remaining on the fruit and leaves becomes a new source of exposure. Initially, the concentrations of these residues are high. They decline over time as a result of biodegradation and exposure to light. The rate of dissipation of foliar residues varies considerably with different pesticides. Weather also affects the rate of dissipation. Rain removes pesticides more rapidly, and high temperatures can change some pesticides to more toxic forms. Too early an entry to a treated site creates a hazard for the worker. Finally, the children of workers are exposed both directly in the fields, where they may be brought when too small to be left alone, and at home, where they are exposed indirectly from the clothing and skin of workers.

2.3.1 Weeding and Harvesting

Workers are at risk during thinning, cultivating, irrigating, and hand-harvesting the crop. One type of worker at special risk is a "pest scout", who enters the fields regularly to count the number of pests on the plants. Since the job often involves going into the field shortly after application, there is a special risk of residue intoxication. This type of illness is sometimes called "picker poisoning." It occurs most frequently with exposure to plants with large leaf surfaces, such as citrus, peaches, grapes, and tobacco. Workers in the field are also at risk of inadvertent direct spraying by aerial applicators.

Exposure varies with different crops. One study indicates that among all crops, citrus represents the "greatest potential hazard" for reentry poisoning to farmworkers. Unlike citrus tree pruners, cultivators, and irrigators, who are not normally exposed to hazardous amounts of organopesticides, the citrus tree picker is often in immediate and sustained contact with pesticide residues for eight hours a day, six days a week, for a period of several months a year. The citrus picker is at risk from dermal exposure to residues on fruit and foliage and from inhalation exposure of pesticide particles made airborne by the picking operations.[50]

Cotton is another crop that presents special hazards to workers. The cotton-field workers at greatest risk of pesticide exposure are those who: 1. have considerable contact with "dew-wet" cotton foliage; 2. enter fields before the end of the reentry period; and 3. fail to maintain good standards of cleanliness. This is important, since cotton receives more insecticide applications than any other crop in the United States.[51] Crops such as citrus fruits, strawberries, tomatoes and lettuce are very labor intensive. Mechanization has in part taken the place of some farmworkers in the harvesting of some fruits and vegetables, including potatoes, and spinach, processing of cabbage, and pickling of cucumbers. Market produce, however, must still be hand-picked. Among the crops that require a lot of hand labor are carrots, green onions, iceberg lettuce, broccoli, celery, strawberries, melons, and tomatoes.[52] If these fruits and vegetables and their foliage contain significant amounts of pesticide residues, "hands-on" farmworkers may be at significant risk, because harvesting brings workers into direct contact with pesticides. Indeed, dermal absorption appears to be the dominant route for pesticide exposure. Among peach harvesters, dermal absorption accounted for 98 to 99 percent of total absorption.[53]

The characteristics of the pesticide itself affect dermal absorption. Lipid-soluble pesticides (such as parathion, DDT, aldrin, and toxaphene) are absorbed more easily than water-soluble pesticides. Reportedly, those pesticides with some solubility in both lipids and water produce the most rapid and complete dermal absorption.[54] One study simulated field conditions through carefully controlled experiments to assess the absorption of pesticides through human skin tissue. Fixed amounts of different pesticides were applied to the skin of subjects' forearms and then compared for absorbency. Proportions absorbed into normal human tissue varied from only 0.4 percent for Diquat to 73.9 percent for Carbaryl.

The results of this study also highlight the importance of dermal absorption. Since tissue of the forearm is relatively resistant to absorption compared to almost all other areas of the body, including the hand, scalp, and forehead, these percentages of absorption are probably minimal figures. In addition, the study found that damaged skin or occluded skin (such as a hand covered by a glove) absorbs much more of the chemical. For example, damaged skin tissue in the forearm absorbs about eight times as much parathion as does normal tissue -- 73.2 percent versus 8.6 percent. Also, occluded normal skin absorbs more than six times as much parathion as does normal (nonoccluded) tissue.[55] Physical movement and sweating have also been associated with increased dermal absorption of pesticides.[56] Specific crop-risk studies of Texas farmworkers could be very informative. Factors to be investigated would include the age and sex distribution of work crews, apparel, and foliar contact and exposure. One such study, for example, found that the orange harvest was conducted disproportionately by women and children under seven years of age. The same study found that the more experienced fieldworkers almost "invariably wore long sleeves." Finally, orange pickers wear gloves, but peach pickers rarely do.[57] These kinds of findings would assist in assessing the effects of some kinds of pesticide exposures and in adopting appropriate policies to forestall these exposures.

2.3.2 Children and Families

Pesticides also pose a risk for family members of agricultural workers. Young children often play in the fields where their parents are working because other childcare alternatives are expensive or unavailable. Consequently, children are at the same risk of residue intoxification as the workers themselves (or possibly more so, due to lack of protective clothing and the fact that they are closer to the ground). Children are also at increased risk of accidental poisonings, from pesticides which have not been stored or disposed of properly, or from direct ingestion through eating contaminated food or soil. Those suffering from pica, a common malnutrition disease of children that causes them to eat dirt, may be at particularly high risk.

Exposures of small children may be especially serious because a given dose of a pesticide is relatively more poisonous to a small child than to an adult. Children are susceptible both because of their lower body weight and because their bodies are still developing. Growing tissue is more susceptible to toxins. The organs of children eight years of age are actually able to detoxify chemicals more efficiently than those of adults. The organs of younger children, however, are not as efficient. For instance, it has been reported that the liver of an infant is seven times slower than that of an adult in the detoxification of certain chemicals. Moreover, it has been argued that the brain, nervous system, immune system, and liver of an infant are particularly vulnerable to toxics in the environment. Children with nutritional deficiencies may be at an even greater risk, since vitamin and mineral deficiencies apparently reduce resistance to the effects of some carcinogenic pesticides such as DDT.[58]

Again, some studies have rather different findings. One study found that for strawberries children are not exposed to greater amounts of pesticide residues. Residues were collected on patches worn inside clothing. It was found that children's patches collected less residue on an hourly basis than did those of adults. This may be because children are less diligent in picking the berries. Since the patches of children and adults collect about equal amounts of residues as a proportion of body weight, however, any additional susceptibility of children will result in increased exposure.[59]

Another researcher has also concluded that the "worst case exposure for children working in harvesting operations in fields and orchards would be roughly that of adults in the same circumstances." Again, the study finds that children might receive a higher absorbed dose per equivalent exposure due to differences in body weight and (possibly) greater absorption potential. The key point is that differential risk to children is probably more a product of toxicological susceptibility than of "exposure related parameters."[60] Children almost certainly have "a greater risk of developing chronic or delayed toxic effects than adults who sustain the same exposure."[61]

It should be noted that EPA pesticide-residue standards do not reflect the different susceptibilities of children. First, the agency sets acceptable exposure levels using an "average diet of a 132 pound male teenager," disregarding the low body weight and limited diet of children. Second, acceptable levels for children are arbitrarily determined by applying a factor of 10 or 100 to presumed safe adult levels, rather than being based on studies of children. For example, EPA (unlike FDA) uses only mature animals in testing pesticides.[62]

Spouses, older children and other family members not in the field also face a risk of exposure transmitted from the fieldworkers themselves. Direct physical contact with a contaminated worker's skin or clothing (or that of a contaminated "field child"), and even laundering contaminated worker (or child) clothing with other household laundry may result in exposures to nonfieldworkers.

2.3.3 Other Factors

As discussed in section 2.2.1.1, hypersensitivity to certain pesticides can represent a special risk for farmworkers who have been exposed over a long period of time. There are certain subpopulations which are more sensitive to the effects of low-dosage environmental pollutants. General factors that enhance sensitivity include developmental periods, genetic conditions, nutritional deficiencies, predisposing diseases, and personal habits. For instance, a deficiency of protein in the diet may contribute to hypersusceptibility to DDT and other insecticides. Careful research would be required to quantify the extent of risk assumed by various subpopulations such as Hispanics, women, or the poorly nourished.[63]

2.3.4 Problems in Assessing Risks from Pesticides

Many of the reports discussed are inconclusive. Studies of acute effects suffer from inadequate basic knowledge of symptoms of exposure and the possibility of confusing them with symptoms of other causes. People's inabilities to report what pesticides they used, where they used them, and the other conditions of use only exacerbate the lack of formal requirements that health officials report pesticide incidents.

Studies of long-term effects have additional problems. First, ecologic studies are subject to the influence of confounding factors, and results may not reflect the true relationship between exposure and outcome and the individual level.[64] Such factors as age, race, cigarette smoking, and alcohol use can be complicating variables that may relate to both the hypothesized exposure and the observed adverse effects, making interpretation of associations between pesticide exposure and such adverse effects difficult.

A second problem in such studies is that pesticide exposure itself may be variable. It may be possible to classify agricultural workers as "exposed" or "unexposed," but the extent and type (which chemicals) of exposure are generally unknown. Unfortunately, these may be crucial factors when measuring long-term effects of pesticide use.

Another problem with most epidemiological studies is that dose data are lacking. Measures frequently depend on "yes/no" or "lots/not much/hardly any" distinctions. This is critical. An observed (and measured) result of some exposure means little if the degree of exposure itself was not measured. On this basis no prediction can be made with respect to what amount of "allowable" dose might be expected to effect some acceptable result.[65]

Laboratory studies, which are also used to detect chronic effects, also have drawbacks: 1. it is not known if humans are more or less sensitive than experimental animals to particular carcinogens; 2. the mathematics of extrapolation allow sizeable variations in response estimates; and 3. humans are much more genetically heterogeneous than laboratory mice and are typically exposed to more environmental hazards.[66]

Another problem lies in the statistical limitations of some studies. If the exposed group is small in number, the statistical power of the study may be low. That is, the probability of detecting even modest associations that may exist is limited, making interpretation of negative findings difficult.[67]

Even though such problems limit the conclusiveness of the research, these studies suggest that the deleterious effects of chronic pesticide exposure must be taken seriously. To ignore them or stress their inconclusiveness risks understating the long-term hazardous nature of pesticides, just as interpreting the studies as definitive would risk overstating the problem. Epidemiological and laboratory studies cannot alone produce risk estimates. They must be linked to biological models derived from other research. Useful epidemiologic work in this area will have to adhere to rigorous standards. While this research is being conducted, however, the field studies will have to stand as evidence that pesticides are implicated in both acute and chronic health effects.

3. PESTICIDES AND WORKER HEALTH IN TEXAS

3.1 FARMWORKERS IN TEXAS

Perhaps the outstanding feature of the study of farmworker health is the difficulty of obtaining good information and data. This is true not only in technically complex areas such as inferring a causal relationship between pesticide exposure and health effects, but in areas that would appear to be more straightforward, including counting numbers of workers.

For example, the Texas Crop and Livestock Reporting Service reported that the total number of farm and ranch workers decreased from 219,000 in 1982 to 205,000 in 1983. This is logical, since the number of farms in Texas decreased from about 186,000 in 1981 to approximately 184,000 in 1983, with a concomitant reduction in farmland of about 200,000 acres overall.[68] However, in the same period, the number of hired workers increased from 70,000 (32 percent of the total workers) to 85,000 (41 percent). (The balance of the ranch and farm work force includes the ranch and farm owners and some unpaid persons -- mainly members of owners' families.) About 40,000 (47 percent) of the hired workers in 1983 worked on a seasonal basis -- less than five months.[69] It is not clear how this 1983 figure of 85,000 hired farmworkers can be compared to the 1976 figure of 400,000 persons in migrant and seasonal farmworker households.

In 1976, the (now defunct) Governor's Office of Migrant Affairs distinguished migrant farmworkers from seasonal farmworkers. An individual who had done farmwork within the last five years and had left his or her residence to secure that work was considered to be a migrant farmworker. By contrast, a seasonal farmworker was defined to be any nonmigrant farmworker who worked five months or less per year. About 43 percent of nonmigrant farmworkers are estimated to work on a seasonal basis.[70]

In 1976, a survey estimated that there were about 375,000 persons in migrant households and 120,000 in seasonal farmworker households in Texas. The typical migrant or seasonal farmworker household had six members. Virtually all (95 percent) of the migrant farmworker households were headed by a Mexican (born in Mexico) or Mexican American (born in the United States). Slightly more than one-third (36 percent) of migrant farmworkers were sixteen years of age or younger; only about 1 percent were over sixty-five years of age.

Interestingly, nearly 80 percent of these migrant farmworkers reside in urban settings (cities with populations of 10,000 or more). This is not unlike the nonfarmworker population in Texas. The balance of migrant farmworkers live on farms or in smaller cities. Perhaps surprisingly, about one-half (53 percent) of Texas migrant farmworkers reportedly are homeowners. Only 5 percent were tenant farmers.[71] In addition, it has been estimated that nearly one-half (45 percent) of all Texas farmworkers are female. About 36 percent of these women are heads of households.[72] These sociodemographic characteristics of Texas migrant farmworkers are summarized in table 3.

One of the key problems in assessing the health effects of pesticide exposure on farmworkers is the fact that migrant farmworkers are very mobile and may come into contact with any number of pesticides within Texas and in trips to other states over the course of a year. As table 4 indicates, slightly more than one-third (37 percent) of migrant farmworkers who reside in Texas during the "off-season" report working in some other part of Texas. The

Table 3. Sociodemographic Summary of Migrant Farmworkers
          in Texas, 1976

Age Distribution	Percent
0-5 years	11
6-16 years	25
17-25 years	17
26-35 years	14
36-45 years	15
46-55 years	14
56-65 years	3
66+ years	1

Ethnicity: Head of Household	
Mexican American	72
Mexican	23
Black	2
Anglo	2
Other	1

Residential Setting	
Urban (10,000+)	78
Rural, Farm	11
Rural, Nonfarm	11

Type of Residence	
Owner occupied home	53
Rented house, private	17
Rented apartment, private	13
Living with relatives	7
Tenant farmer	5
Rented house, public	3
Rented apartment, public	2

Source: *Migrant and Seasonal Farm Workers Population Survey:
         Final Report*, July 15, 1976, pp. 27-29.

greatest flow is from the Lower Rio Grande Valley to the Panhandle and, to a lesser extent, to the Middle Rio Grande Valley.[73] Most of the farmworkers in Texas reside in the Rio Grande Valley. The eleven counties with the greatest populations of migrant seasonal farmworkers are listed in table 5. A great many farmworkers live in South Texas, especially in Hidalgo and Cameron counties. However, the Panhandle also has a large share of the state's farmworker population.

The area of Texas in which migrant farmworkers "reside" seems to have a bearing on which states they visit. Migrant farmworkers who travel to other states and reside in far West Texas and the Upper Rio Grande most often visit New Mexico. Migrant farmworkers residing in the Lower Rio Grande Valley tend to frequent the upper Midwest, while Coastal Bend residents tend to favor the eastern Midwest states. Finally, East Texas and Central Texas migrant farmworkers often seek work in southern Florida and Louisiana. In addition, many thousands of documented and undocumented workers from Mexico cross the border regularly to work in the fields.[74] Most of the interested parties agree that the U.S. fruit and industry work force (estimated to be between 700,000 and 1,000,000 persons) is comprised largely of Mexican nationals, many of whom are in the United States illegally. It is claimed that many longstanding networks exist through which additional farmworkers can be quickly delivered to U.S. farmers on short notice.[75]

3.1.1 The Health of Texas Farmworkers

Texas Rural Legal Aid, Inc., cites figures that the average life expectancy of the farmworker is about twenty years less than that of the average American. Infant mortality in farmworker housholds is 24 percent above the nationwide rate. Farmworkers contract influenza and pneumonia 20 percent more

Table 4. Destination of Migrant Farmworkers Residing in Texas

State	percent
Texas	37
New Mexico	10
Ohio	8
Michigan	7
Florida	5
Colorado	4
Indiana	3
California	2
Arizona	2
Illinois	2
Mississippi	2
Nebraska	2
Minnesota	2
Other States	14

Source: *Migrant and Seasonal Farm Worker Population Survey: Final Report*, July 15, 1976, p. 43.

37

Table 5. Texas Counties with Greatest Numbers of Migrant and
Seasonal Farmworkers, 1978

County (HSA)	Total Population	Peak Migrant Population	Seasonal Population
Bexar (IX)	16,140	15,500	640
Cameron (VIII)	59,300	53,000	6,300
Deaf Smith (I)	12,000	10,000	2,000
Hale (II)	12,300	9,700	2,600
Hidalgo (VIII)	77,000	63,000	14,000
Lubbock (II)	10,900	8,000	2,900
Maverick (IX)	10,560	10,000	560
Nueces (VIII)	8,500	6,500	2,000
Parmer (I)	11,800	11,000	800
Starr (VIII)	8,500	6,700	1,800
Webb (VIII)	21,000	18,000	3,000
Statewide	482,595	373,495	109,100
Statewide (Adjusted)	430,000	294,000	136,000

Source: U.S. Department of Health and Human Services,
1978 Migrant Health Program Target Populations Estimates
(April 1980), pp. 85-91.

often than does the typical nonfarmworker.[76]

Unsanitary drinking water and improper waste water disposal can contribute to poor health status. Bacterial disease is easily transmitted, and heat stroke, dehydration, and other acute symptoms are exacerbated by lack of good water and toilet facilities. The section on pesticide exposure also shows that availability of water in which to wash is a critical feature in reducing many acute symptoms of pesticide exposure. Unfortunately, in the past the costs of these facilities discouraged some farmers from providing them.

In June 1983, the Texas Department of Health adopted regulations to upgrade and maintain healthy working conditions for farmworkers in Texas. The regulations establish standards for the provision of on-site toilet, handwashing, and drinking water facilities for all those who employ more than six people on any workday at a temporary place of employment. These facilities may be portable. Each handwashing facility is required to have "a suitable cleansing agent" and individual hand towels. If the supply of potable water is insufficient for handwashing purposes, "disposable, pre-moistened cleaning towels and emulsifiable skin cleaners" may be substituted. In addition, the regulations prohibit the storage, preparation, or consumption of food in any area "where there are materials or substances present in quantities or concentrations which may contaminate food or be injurious to health."[77]

According to Troy Lowry, registered sanitarian with the TDH Migrant Labor Housing Sanitation Branch, it is estimated that less than 10 percent of the growers in the Valley provided any such facilities before the regulations were drafted. Since then, however, about 85-90 percent of the Valley growers are believed to be in compliance. Local health departments have the primary enforcement responsibility. Informal enforcement began in the Valley first and is now underway in the Panhandle. Written warnings for violations will probably not be issued until the 1984 summer growing season begins. TDH reports good cooperation from the growers through the Texas Citrus and Vegetable Shippers Association. When requested to do so, TDH provides information on the availability of manufactured portable facilities. Approximately 500 such units have been purchased for agricultural worksites in the last six months.[78]

3.1.2 Rural Health Care

One problem heightening any potential health problem among farmworkers in Texas is the relative lack of facilities and practitioners and inadequate transportation to health care providers. For instance, one study indicated that among thirty-two counties in South Texas in which farmworkers reside and are employed in great numbers in 1978:

> Twenty-three (72 percent) counties had no public health department;
> Twenty (63 percent) had no public hospital;
> Twelve (38 percent) had no emergency room;
> Nine (28 percent) had no hospital;
> Four(13 percent) had no ambulance.[79]

There is also a serious shortage in Texas of Hispanic physicians who might better bridge the cultural and language gap between the farmworker and the health care provider.[80] In Texas, migrant and seasonal farmworkers typically do not seek or receive needed medical services because they are generally not able to take time off from work.

When clinic facilities are geographically accessible, they often operate only during the day time and many times treat only residents.[81] Data on specific kinds of facilities are provided in tables 6 to 8. One indicator of the availability of care is the number of hospitals and beds; table 6 gives data for nineteen counties with high numbers of agricultural workers and/or high agricultural production. Another indicator of health care availability is the number of doctors. According to the American Academy of Family Physicians (AAFP), the preferred ratio of population to primary care physicians is 2,000:1. The federal government has set standards which establish an upper limit of 3,500:1 or 3,000:1 if the area has "high need." Areas exceeding these limits are designated as primary care physician manpower shortage areas.[82] Based on 1982 data, shown in table 7, it can be seen that statewide, Texas had a moderate surplus of primary care physicians with a ratio of 1,781 persons per primary care physician. However, an analysis of the same nineteen counties reveals that twelve of them exceed the AAFP ratio -- from 2,059:1 in Hansford County to 11,334:1 in Randall County.

Of particular importance is that seven of the nineteen counties -- Deaf Smith, Parmer, Castro, Randall, Webb, Maverick, and Starr -- in 1982 had population/primary care physician ratios in excess of 3,500:1. Interestingly, there were enough primary physicians among the nineteen counties to provide a ratio of 1,935:1 in each county if the physicians had been allocated on a strict population basis.

Over the past decade or so, there has been an increase in the number of physicians practicing in rural counties over the preceding ten year period.[83] Within the state, the Texas Academy of Family Physicians has implemented a program which facilitates meetings between family practice residents and communities in need of primary care physicians.[84]

However, between 1981 and 1982, six of the nineteen agriculture/farmworker counties realized no gain in primary care physicians. All six of these counties -- Deaf Smith, Parmer, Castro, Sherman, Swisher, and Randall -- have population/primary care physician ratios in excess of 3,000:1.

Table 6 provides similar data for "total physicians," or primary care plus all other physicians. In sum, sixteen of nineteen counties exceed the statewide ratio of 513 persons per physician. For eight counties the total number of physicians did not change between 1981 and 1982, and one county lost two physicians.

Table 6. Number of Short Term Care General and Special Hospitals
and Hospital Beds in Top Agricultural/Farmworker Counties

| County | # Hospitals | Number of Hospital Beds | | | # Conform. Beds Per 1000 Pop. (1982) |
		Licensed	Operating	conf	
Deaf Smith	1	77	68	70	3.16
Parmer	1	34	34	43	3.79
Castro	1	46	46	56	5.44
Hansford	1	28	28	38	6.15
Hidalgo	6	552	548	559	1.79
Moore	1	80	80	91	5.32
Sherman	0	0	0	0	0.00
Swisher	1	30	30	30	3.06
Lamb	2	110	86	75	3.95
Hale	3	218	221	166	4.29
Randall	1	49	49	48	0.60
Dallam	0	0	0	0	0.00
Lubbock	8	1546	1324	1305	6.02
Cameron	4	616	499	390	1.70
Webb	2	318	307	229	2.11
Bexar	17	4966	4437	4614	4.50
Maverick	1	77	77	77	2.16
Nueces	8	1671	1485	950	3.45
Starr	1	44	44	44	1.45

Notes: Licensed beds = the total number of beds the facility is authorized to operate by the licensing authority. (Records updated through October 31, 1980.)

Operating beds = the number of beds set up and staffed for use with support services.

Conforming beds = the number of beds found through a survey by the Texas Department of Health to be in compliance with certain established standards. (Records updated through October31, 1980.)

Source: *1979 TDH Integrated File*. This list includes all licensed short-term care (under 30 days) nonfederal or state community general and special hospitals which were open and reported utilization data during the reporting period and which were available to the general public (as cited in *The Texas State Health Plan*, adopted by Texas Statewide Health Coordinating Council, March 26, 1982, pp. 600-609).

Table 7. Number of Primary Care Physicians in Top Agricultural/
Farmworker Counties, 1982

Population/Physician

County (HSA)	# Primary Care Physicians	#	County to State Ratio	County to AAFP Stndrd. Ratio
Deaf Smith (1)	6	3689 **	2.07	1.84
Parmer (1)	3	3782 **	2.12	1.89
Castro (1)	2	5463 **	3.07	2.73
Hansford (1)	3	2059 **	1.16	1.03
Hidalgo (8)	115	2713 **	1.52	1.36
Moore (1)	13	1315	0.74	0.66
Sherman (1)	1	3088 **	1.73	1.54
Swisher (1)	3	3264 **	1.83	1.63
Lamb (2)	10	1898 *	1.07	0.95
Hale (2)	27	1432	0.80	0.72
Randall (1)	7	11334 **	6.36	5.67
Dallam (1)	5	1329	0.75	0.66
Lubbock (2)	147	1475	0.83	0.74
Cameron (8)	84	2728 **	1.53	1.36
Webb (8)	29	3748 **	2.10	1.87
Bexar (9)	588	1744	0.98	0.87
Maverick (9)	8	4455 **	2.50	2.23
Nueces (8)	213	1293	0.73	0.65
Starr (8)	6	5068 **	2.85	2.53
Texas (Statewide)	8391	1781	1.00	0.89

Notes: Asterisk indicates a value which exceeds the statewide average; double asterisk indicates a value which also exceeds the AAFP standard.

The American Academy of Family Physicians (AAFP) recommends 2,000:1 as the preferred ratio of population to family physicians.

Source: *Texas Health Manpower Report: Physicians 1982*, TDH, Bureau of State Health Planning and Resource Developments.

Table 8. Number of Physicians (Total) in Top Agricultural/
Farmworker Counties, 1982

			Population/Physician	
County (HSA)	Population	# Physicians	#	County to State Ratio
Deaf Smith (1)	22,134	9	2459 *	4.01
Parmer (1)	11,534	3	3782 *	6.17
Castro (1)	10,925	4	2731 *	4.46
Hansford (1)	6,178	5	1236 *	2.02
Hidalgo (8)	311,966	254	1228 *	2.00
Moore (1)	17,100	15	1140 *	1.86
Sherman (1)	3,088	1	3088 *	5.04
Swisher (1)	9,792	3	3264 *	5.32
Lamb (2)	18,976	14	1355 *	2.21
Hale (2)	38,657	45	859 *	1.40
Randall (1)	79,339	9	8815 *	14.38
Dallam (1)	6,645	5	1329 *	2.17
Lubbock (2)	216,772	502	432	0.70
Cameron (8)	229,139	222	1032 *	1.68
Webb (8)	108,701	78	1394 *	2.27
Bexar (9)	1,025,316	2589	396	0.65
Maverick (9)	35,640	18	1980 *	3.23
Nueces (8)	275,356	529	521	0.85
Starr (8)	30,408	7	4344 *	7.09
Texas (Statewide)	14,944,493	24376	613	1.00

Note: Asterisk indicates a value which exceeds the statewide average.

Source: *Texas Health Manpower Report: Physicians 1982*, TDH, Bureau of State Health Planning and Resource Development.

3.2 EXPOSURE DATA FROM TEXAS

Incidents of acute pesticide poisoning cases in the Rio Grande Valley have been selectively reported since 1960. In that year, Dr. George L. Gallaher opened a Poison Control Center at Valley Baptist Hospital in Harlingen, which maintained records of the annual number of poisonings reported to it. Since physicians were not required to report pesticide poisonings to the Center, and since many mild poisonings are never medically treated, the Center's figures may grossly understate the actual numbers. Data sources other than the Poison Control Center (and Dr. Gallaher) were not introduced until 1968.

According to the Center's records, medical care was obtained for about 275 pesticide poisoning cases in the Valley from 1960 through 1966. About 25 percent of these cases took place during the first four years of that period. In 1964, there was a striking increase in the number of cases; the total (70) was approximately equal to the number of cases observed during the previous four years combined.[85] This rise "may have been the result of the estimated 400 percent increase in pesticide usage during the year, or to the relatively sudden changeover from the less acutely toxic organochlorine compounds, mainly DDT and benzene hexachloride, to the more acutely toxic organophosphate compounds, mainly parathion."[86] In each of the years 1965 and 1966 the Center also treated about 70 cases.

In 1968, the Texas Community Studies Pesticide Project, through a contract with the United States Department of Health, Education and Welfare, began documenting pesticide poisoning cases in the Valley. Data sources included annual hospital inpatient records of the six major hospitals, direct contact with all pesticide formulating and application companies, and personal communication with several physicians, as well as Poison Control Center data. In 1968, the Texas Community Study documented 118 cases in the Valley, 51 of which were treated at the Poison Control Center. Poisonings were most frequent during the summer, during the peak formulation period and early application period. Aerial applicators accounted for 62 percent and growers and farmworkers accounted for 28 percent of the total cases in 1968. A mass poisoning in one field accounted for most of the grower and farmworker cases.

Incidence was highest among teenage boys.[87]

In 1969, the Texas Community Study documented only 15 cases (no Poison Control Center data). "There are several reasons for the decrease. Pesticides were applied less frequently in 1969, and the most widely used product was less toxic than that of the previous year. Also, better safety habits were followed."[88]

From 1971 through 1976, Texas data are available from the *National Study of Hospital Admitted Pesticide Poisonings*. During that period, the number of hospital admitted poisonings decreased in the state. In 1971, there were an estimated 355 hospital cases; in 1976, the number had dropped to 256. Correspondingly, the rate per 100,000 hospital admissions decreased from 17.0 to 11.3.[89] It should be emphasized that this trend cannot necessarily be generalized to represent all acute poisonings in Texas. The number of poisonings treated outside a hospital and the number not treated at all are simply not known.

The federally sponsored Pesticide Incidence Monitoring System (PIMS) began in Texas in 1977 in San Benito. By 1979, the Southeast Texas Poison Center in Galveston was getting 900-1,000 calls a year under the PIMS system.[90] Those statistics were sent to the Texas Tech lab at San Benito for collating. In 1980 a total of 2,335 calls were recorded from the five PIMS centers reporting to the San Benito lab.[91] About 60.5 percent of the reports came from centers in Texas, although it may not be accurate to separate the figures by state.

In April 1980, a pesticide forum was convened in Pharr, Texas by the

National Rural Health Council of Rural America. The forum was attended by farmers and workers as well as representatives of all other interested groups, including manufacturers and officials from several levels of government.[92] The forum group did not attempt to estimate the number of poisoning cases occurring in the state. However, despite fears of retaliation, 40 affidavits from farmworkers themselves were formally submitted into the record, and together they represent further documentation of a pesticide-related health problem in the Valley.

These affidavits, field studies conducted since 1980, and the records of several major pesticide exposure incidents in the state all confirm that Texans suffer in the same ways as other agricultural workers. Dermatitis, headaches, dizziness, and eye problems are widely reported. At the Pharr conference, for example, the health complaints most commonly mentioned due to pesticide exposure were rashes and sores on the skin, headaches, and eye irritation. Other conditions detailed include nose bleeds; swollen hands, arms, faces, and eyes; vomiting; dizziness; difficulty in breathing; miscarriages; and chest pains. Two cases of permanent disability, loss of eyesight and amputation of a foot due to infection, were also described.

Similarly, preliminary analysis of a survey conducted in 1981 by the National Association of Farmworkers Organizations "found that 80 percent were experiencing dermatitis linked to pesticide exposure, 40 percent suffered from chronic headaches, 51 percent complained of dizziness and 23 percent reported blurred vision. In a study funded jointly by the EPA and the National Academy of Science, 56 percent of the agricultural workers surveyed had abnormal liver and kidney functions, 78 percent reported chronic skin rash and 54 percent had chest cavity abnormalities."[93]

Similar symptoms have been documented in several major pesticide exposure incidents. These incidents are important, since they provide the most direct evidence of a causal relationship between pesticide exposure and health effects. In 1968, 23 people who had been working for about 3 hours in a dew-wet cotton field that had been sprayed 12 hours before with methyl and ethyl parathion were treated for pesticide poisoning at the Harlingen Poison Control Center. Of the 23, 13 had to be hospitalized and 10 were treated on an outpatient basis. A similar incident occurred in 1971, when the same pesticides had been sprayed on a cornfield. Over a period of 48 hours, 36 of the farmworkers (23 percent) reported to the Uvalde Memorial Hospital for poisoning. One-half of these individuals were hospitalized; the other half received emergency treatment and were released. Interestingly, all of those treated and released or hospitalized and discharged by June 14 returned to work that day.[94] This may be an indication of the farmworkers' belief that pesticides present little real danger to their health or of the pressing need to take available work opportunities.

Pesticide exposure in Texas occurs in the ways described above. Among the factors contributing to exposure are:

1. use of many different pesticides;

2. relatively frequent spraying of farmworkers by aerial applicators;

3. lack of adequate protective clothing;

4. improper disposal of empty pesticide containers;

5. inadequate or ineffective posting of warning signs for fields that have been treated;

6. lack of training and education of the applicators, farmers, fieldworkers, inspection and enforcement officials, and medical professionals;

7. lack of childcare alternatives (so children are in the fields);

8. equipment malfunctions;

9. lack of appropriate toilet, bathing, and drinking water facilities;

10. lack of monitoring of pesticides of persons who work with them;

11. an inadequate complaint process including the absence of feedback on complaints submitted.[95]

Again, the testimony at the Pharr conference summarizes these problems: "The majority of the farmworkers who testified have no idea what chemicals they have been in contact with and were not informed when the fields had been sprayed or were not warned when spraying was to take place." Of the 11 who explained how they were exposed, 6 mention being directly sprayed, 3 came in contact with drift, and 2 experienced problems related to reentry into a treated field.[96]

3.3 AN EXPLORATORY ANALYSIS OF TEXAS DATA

In addition to the kinds of evidence presented in section 3.2, another means of assessing the relationship between pesticides and worker health is to correlate morbidity and mortality data with pesticide exposure data. Such a study would be subject to many of the same problems of scientific interpretation that we have found with other aspects of the exposure-health relationship. For example, we noted above that misdiagnosis of pesticide poisoning is likely. Symptoms of pesticide poisoning are similar to those of brain hemorrhage, heat stroke, heat exhaustion, hypoglycemia, gastroenteritis, pneumonia or other severe respiratory infection, and asthma.[97] If pesticide poisoning is mistakenly diagnosed as other illnesses in a systematic fashion, then the "overincidence" of those other illnesses might be indicative of pesticide poisoning. The available data may be of questionable quality on other dimensions as well; it is of particular importance in this regard that no pesticide use data are available for Texas, so that exposure must be measured by indirect indicators such as amount of crop produced.

Because of these difficulties, experts differ even on whether it would be worthwhile to conduct exposure-health studies. Some believe that the quality of the data makes any study a waste of money. Morbidity data are limited and generally unavailable. Mortality data are suspect since there is no consistent, high-quality determination of the cause(s) of death across the state. In many cases, the individual specifying the cause of death has very little medical training. In addition, information may be lost. Assuming that people might be dying from specific types of cancer as a result of exposure to specific types and amounts of pesticides, there is no guarantee that this "fact" would be reflected in the specified cause of death. An individual might be hospitalized for treatment of the tumor, and subsequently develop pnneumonia and die. The recorded cause of death might be pneumonia or some other complication. This information might also be lost if the person should die in an automobile accident.[98]

Ephraim Kahn, a renowned epidemiologist in the field, stated that he does not know how he would go about determining whether or not there was a link between pesticide usage and specific cancers, given the inadequacy of data and complexity of the problem. Kahn notes the relatively long latency period (twenty to thirty years) for carcinomas as the key stumbling block to analysis. A properly conducted study would require researchers to know the types and amounts of pesticides used in given localities twenty to thirty years ago. Moreover, one would need to trace workers back to their work sites over the same time period.[99]

However, other experts feel that a study of causes of mortality by amount of pesticides used in Texas would be worthwhile. Although available data would not permit us to infer clear cause-and-effect relationships between pesticides and specific health effects, an understanding of the causes of death for agricultural workers could form the basis for more formal studies.[100]

One possible way to circumvent the poverty of aggregate data is to pinpoint those Texas counties which are "hot spots" for relevant cancers, and then to analyze in detail pesticide usage and the incidence of poisoning in one or two of those counties.[101] This approach has been recommended in the 1982 State Plan for Texas Farmworkers, which suggests that the Texas Department of Health should consider expanding its Cancer Program to "pinpoint geographic areas of high pesticide use" while monitoring cancer rates among farmworkers.[102]

Keeping these problems in mind, we decided to make use of existing data to conduct a preliminary study of pesticides and worker health in Texas. This section attempts to compare a proxy measure of pesticide usage data with cancer morbidity and mortality data on a county-by-county basis for the 254 counties in Texas. Pesticide usage data are unavailable in Texas, since pesticide manufacturers prefer not to make this information public. In lieu of these data, Clark et al. used the proportion of the total land area in cropland in each county as a proxy measure of pesticide use in their study exploring the link between agricultural activity and cancer mortality in the southeastern United States.[103] This proxy measure assumes that most harvested cropland is treated with pesticides. We also use this measure, and also consider counties with high proportions of agricultural workers, on the assumption that these are counties where pesticide use and consequent exposure is high.

The underlying assumption of any study using aggregate data is that occupational exposure does vary significantly among counties and that agricultural counties with farm and rural populations have a greater net risk than nonagricultural counties. One major criticism of this assumption which cannot be adequately addressed within the present analysis is that urban counties may have any number of nonpesticide-related factors which affect mortality and therefore confound any direct comparisons of mortality rates. Cancer of the liver may be caused by chronic exposure in the field but it may also be caused (to a greater extent) by the stress of living in a large city.

3.3.1 An Analysis of Cancer Morbidity Data

Drs. Macdonald and Heinze of The University of Texas System Cancer Center's M. D. Anderson Hospital and Tumor Institute have compiled an extensive dataset of the incidence of cancer in selected counties of Texas,[104] representing nearly 160,000 cases of cancer in residents of 56 counties during the period from January 1944 through December 1966. The data, which are probably as complete as possible with present record-keeping methods, were grouped on a regional basis and age-adjusted for each of 37 anatomical sites of cancer. We correlated these data with the proxy pesticide usage data -- the proportion of total land area devoted to harvested cropland in 1945. The proportion figures were calculated based on Census of Agriculture data. The 17- to 20-year difference between the cancer morbidity data and the crop/pesticide intensity data was considered appropriate in light of the latency period of cancer.

The six regions defined by Macdonald and Heinze encompass only the southern
and southwestern portions of the state, and can be roughly compared to the
Health Service Areas (HSAs) used in the analysis of the TDH data below and
shown there in figure 1. The Corpus Christi, Harlingen, and Laredo regions
taken together closely approximate HSA VIII. The San Antonio region includes
a few more counties than does HSA IX. The El Paso region is the same as HSA
III. The Houston region (Harris County) is at the center of HSA XI and is the
only area not contiguous with at least one other region. These six regions
vary considerably in the amount of total land area and in the number of
counties covered, which varies from one (Harris) in the Houston region to
twenty-eight in the San Antonio region. Table 9 provides the detailed 1945
harvested cropland data used to calculate the proxy usage variables. Region
wide, crop intensity was relatively low in 1945, with only about 6 percent of
the available land used for crops compared to the statewide figure of about 16
percent. Since crop intensity varies a great deal within the regions,
especially the larger regions such as San Antonio and Corpus Christi,
information is simply lost in the formulation of aggregate values.

Table 10 lists the average age-adjusted incidence rates for ten sites of
cancer which have been potentially linked to pesticide exposure in one or
major studies. The total cancer rate (for all sites of cancer) is also
listed. It should be pointed out that these rates are aggregated by the
county of residence of the patient and not by the place of diagnosis or
treatment. Rates are expressed as the number of cases per 100,000 population.
These rates should not be confused with mortality rates. Rather, they

Table 9. Percent of Land Area Devoted to Harvested Cropland in Counties for Which Cancer-Incidence Data Exist (Based on Macdonald and Heinze Study)

County (Region)	1945 Total Cropland Harvested (TCH) in Acres	1954 Total Land Area in Acres	Percent of TLA in Region Which is TCH
(Corpus Christi Region)			
Aransas	3,428	176,640	
Bee	78,726	538,880	
Calhoun	28,641	343,680	
Goliad	43,008	557,440	
Kenedy	90	900,480	
Kleberg	26,535	544,640	
Live Oak	72,296	686,080	
Nueces	252,055	536,320	
Refugio	35,383	493,440	
San Patricio	205,891	440,960	
Victoria	67,928	571,520	
Corpus Christi Region Total	813,981	5,790,080	14.058
(El Paso Region)			
Brewster	747	3,973,120	
Culberson	1,796*	2,462,720	
El Paso	56,847	674,560	
Hudspeth	14,959	2,901,120	
Jeff Davis	1,168	1,445,120	
Presidio	5,769	2,481,280	
El Paso Region Total	81,286	13,937,920	0.583
(Harlingen Region)			
Cameron	122,239	565,120	
Hidalgo	205,883	986,240	
Starr	39,991	772,480	
Willacy	92,452	380,800	
Harlingen Region Total	460,565	2,704,640	17.029
(Houston Region)			
Harris	82,755	1,107,200	
Houston Region Total	82,755	1,107,200	7.474
Brooks	15,197	578,560	
Duval	50,908	1,160,960	
Jim Hogg	9,385	731,520	
Jim Wells	109,040	541,440	
Webb	18,753	2,108,800	
Zapata	6,195	691,200	
Laredo Region Total	209,478	5,812,480	3.604
(San Antonio Region)			
Atascosa	139,745	423,680	
Bandera	23,922	489,600	
Bexar	163,563	798,080	
Blanco	25,467	460,160	
Comal	31,554	362,880	
De Witt	117,597	582,400	
Dimmit	14,919	858,240	
Edwards	2,558	1,328,000	
Frio	68,346	714,240	
Gillespie	81,369	675,200	
Gonzales	106,668	677,120	
Guadalupe	158,576	457,600	
Karnes	182,901	485,120	
Kendall	29,637	428,800	
Kerr	34,769	704,640	
Kimble	16,198	815,360	
Kinney	4,105	890,240	
La Salle	46,694	960,640	
Lavaca	119,867	624,000	
McMullen	8,245	741,760	
Maverick	18,432	818,560	
Medina	99,734	865,920	
Real	4,002	400,000	
Terrell	738*	1,528,320	
Uvalde	33,283	1,016,320	
Val Verde	2,688	2,074,880	
Wilson	126,206	513,280	
Zavala	25,803	826,880	
San Antonio Region Total	1,687,586	21,521,920	7.841
Six Region Total	3,335,651	51,874,240	6.430
Texas Total	27,469,089	168,648,320	16.288

*1945 data not reported; 1949 data substituted.

Sources: U.S. Bureau of the Census, Census of Agriculture (Texas): 1945 Vol. 1, Part 26, pp. 347-359; and Census of Agriculture (Texas), 1954 Vol 1 Part 26 pp 77-83

represent the total number of persons (without regard to race, sex, or age) who present themselves to a health care provider and are subsequently diagnosed to have a malignant tumor. The data were also age-adjusted against the standard 1960 U.S. population.

From table 10 it can be seen that the Corpus Christi and Houston regions have the highest cancer morbidity rates relative to the other four regions. (Corpus Christi with a crop intensity measure of 14 percent is about average compared to the state as a whole while Houston at 7 percent is below average.) Corpus Christi in particular has very high rates. Seven of the ten specific site rates exceed those of the six region aggregate. The cancer of the skin rate, 187, is more than double that of the aggregate, 82. It is also more than double that of the El Paso region, 89, which is surprising, given the fact that El Paso has the highest amount of available sunlight. Increased exposure to sunlight has of course been linked with increased skin cancer.[105] Corpus Christi also has the highest rate of soft tissue sarcoma.

Houston also exceeds the aggregate rates for seven of the ten specific sites and for the total rate. Of the six regions, Houston has the highest rates for cancers of the brain and central nervous system, leukemia, lung, prostate, and upper respiratory system. Its prostate rate, 51, is exceptionally high; Corpus Christi is second with a rate of 38. By contrast, Houston has a very low rate of skin cancer, 47, which is roughly one-half the aggregate rate.

There are a number of factors quite distinct from pesticide usage which could easily explain the high cancer morbidity rates in the Houston and Corpus Christi regions. First, these areas are major centers for the petrochemical industries, which may very well present increased risks not only to those persons who work with hazardous petrochemicals but also to those in the larger community who come into contact with pollutants in the environment. Second, many households in the Gulf Coast and South Texas regions rely on groundwater for their drinking water supply. Untreated water supplies generally increase the chance for passing biological and chemical contaminants along to the population. Finally, the natural concentration of uranium is very high in these regions. Several counties have open-pit uranium mines. In some areas uranium has seeped into the groundwater. The ability of radioactive materials to induce mutations has long been accepted.

Harlingen, the most crop-intensive region (17 percent), deserves attention. Its incidence of stomach cancer is the highest but only slightly so. Skin cancer is quite prevalent; its rate of 107 is second only to the extraordinary rate of Corpus Christi. Four of Harlingen's ten specific site rates are above the aggregate rate. Its total cancer rate, 365, is virtually identical to the aggregate rate, 366, and -- curiously enough -- is very similar to the Houston total rate, 368. That two such disparate regions should yield the same overall rate of cancer is striking.

Table 11 presents the results of a simple correlation between crop intensity and cancer morbidity across all six regions for each of the ten specific cancer sites and for total cancer. For three of the sites -- leukemia, liver, and upper respiratory -- the correlation coefficients are virtually zero. That is, there is no correlation between the two variables. Perfectly correlated relationships yield coefficient values of +1 or -1 depending on whether the association is positive (in this case, whether the likelihood of

Table 10. Average Age-Adjusted Rate of Incidence of Cancer per 100,000
Residents by Site of Cancer for Selected Regions, 1962-66

Site of Cancer	Corpus Christi	El Paso	Harlingen	Houston	Laredo	San Antonio	All Six Regions
Brain and Central Nervous System	4.307	3.591	3.830	4.969*	3.303	4.161	4.382
Leukemia	9.132*	7.764	6.883	10.548*	6.344	8.305	8.927
Liver	3.278*	2.215	3.002*	2.510	5.439*	2.652	2.814
Lung	30.182*	18.717	22.307	32.864*	20.195	22.578	26.893
Lymphoma	9.995*	6.628	6.829	9.140*	6.835	8.394*	8.367
Prostate	38.414	23.364	31.996	51.281*	34.365	35.334	39.727
Skin, Exposed Areas	186.649*	88.794*	107.227*	46.931	72.702	74.887	81.696
Soft Tissue Sarcoma	3.962*	2.490	2.803*	2.466	2.408	2.635	2.732
Stomach	10.260	10.379	12.113*	12.047*	11.726*	11.523	11.579
Upper Respiratory System	11.939*	6.183	7.048	13.171*	11.592*	8.915	10.539
All Sites	496.397*	321.353	364.534	368.462*	332.216	329.923	366.480

*Asterisk indicates that the rate exceeds the six-region total rate.

Source: Eleanor J. Macdonald and Evelyn Heinze, *Epidemiology of
Cancer in Texas: Incidence Analyzed by Type, Ethnic Group,
and Geographic Location* (New York: Raven Press, 1978).

cancer increases with exposure to pesticides, measured here by the proxy variable total cropland) or negative (cancer decreases with exposure). The strongest suggestion of a relationship occurs for soft tissues, followed by skin cancer with a value of 0.55. It is interesting that the coefficient for total cancer at 0.62 is as great as for soft tissue sarcoma. This suggests that there are other specific sites (not traditionally cited as "pesticide-related") with even stronger correlation values. Also interesting is the fact that all but one of the coefficients are positive, "hinting" (to the extent that the magnitudes of the values allow) that as crop-intensity (pesticide usage/exposure) increases specific and total cancer morbidity increases. However, the correlation values are too low to make any strong inferences.

3.3.2 An Analysis of the Mortality Data

We also analyzed TDH's average age-adjusted cancer mortality data for the period 1969 to 1978. As in the preceding analysis these health outcome data were tested for correlation against crop-intensity (pesticide usage data). To adjust for the fact that the mortality data are more recent than the morbidity data, 1949 (as opposed to 1945) harvested cropland data were used. As with morbidity data, mortality rates are expressed as the number of cases (deaths) per 100,000 population. The principal geographic unit of analysis used by TDH is the Health Service Area (HSA). There are twelve HSAs in Texas; their boundaries are shown in figure 1. Again, the health data present some problems. The Cancer Registry of the Texas Department of Health has been attempting to develop a statewide data base by reviewing and encoding hospital records from 1976 to the present. Reportedly, this incidence data base is nearly complete for HSAs I, III, and IX. HSAs V and VI are about 70 percent complete. The data base for HSA XI, however, is only about 20 percent complete. Of particular concern is the fact that the data for HSA VIII, which

Table 11. Results of Simple Correlation of 1962-66 Average
        Age-Adjusted Rate of Incidence of Cancer Data (Based on
        Macdonald and Heinze Study) on 1945 percentage of Region
        Devoted to Harvested Cropland Data by Site of Cancer

Site of Cancer	Correlation Coefficient
Brain and Central Nervous System	0.32
Leukemia	0.08
Liver	-0.09
Lung	0.41
Lymphoma	0.37
Prostate	0.26
Skin, Exposed Areas	0.55
Soft Tissue Sarcoma	0.62
Stomach	0.22
Upper-Respiratory System	0.06
All Sites	0.62

includes some of the most agriculturally intense counties in the state, are virtually nonexistent. Cooperation between registry and hospital officials has been said to be less than ideal.[106]

The analysis that follows involves five distinct treatments of the data. First, a simple correlation between the two variables of interest is performed within each HSA based on county level data. Only total cancer mortality rates are considered. Second, another simple correlation is performed for the thirty-six "most crop-intensive" counties and the thirty-five "least crop-intensive" counties in Texas. Again, only total cancer mortality rates are considered. Third, a weighted correlation based on relative county populations is recalculated for the thirty-six- and thirty-five-county areas. Finally, some of the counties with high farmworker populations are singled out for study.

3.3.2.1 Simple Correlation within HSAs

The first procedure correlating the percent of total land area devoted to harvested cropland has been calculated and paired with the corresponding mortality rate for each of the 254 counties in Texas. Table 12 summarizes the results of this procedure, grouping the counties by HSA. Note that crop-intensity and mortality data listed for each HSA represent an overall value for the entire HSA, but that the correlation coefficients are based on county-level data.

It is important to note the wide range of variation among the correlation coefficients on both sign and magnitude. The most crop-intensive HSA (II) has a positive value, (+) 0.33, but the next most crop-intensive HSA (I) has a negative value, - 0.14. The magnitudes are rather unremarkable. HSA III has

61

Figure 1. Map of the Twelve Health Service Areas (HSAs) in Texas

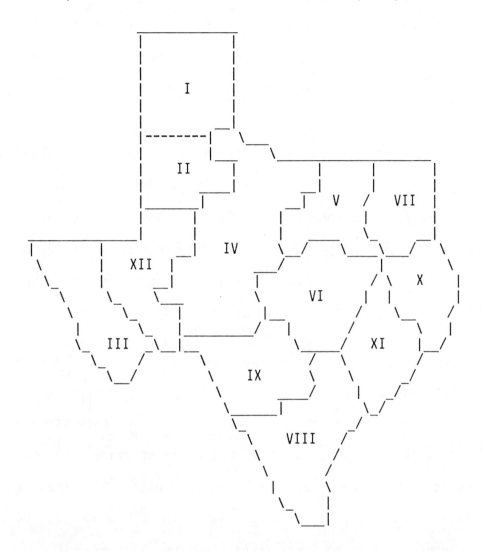

a correlation value of 0.62, but this is almost entirely due to the peculiar data that the area produces. Of the six counties in HSA III, only El Paso County has any appreciable cropland; since it also happens to have the highest mortality rates, the correlation is distorted.

This analysis could be interpreted as an indication that the true relationship between pesticide exposure (as defined within this study) and cancer mortality is nonexistent. Or it could be an indication that any one of several factors that vary across the state and HSAs may be camouflaging an existing relationship. Again, the quality of the health data and the absence of true pesticide exposure data are limiting.

3.3.2.2 Correlation of Extreme Counties

An attempt was made to sharpen the analysis by narrowing the data to include only those counties which are especially crop-intensive or non-crop-intensive. If pesticide exposure is an important cause of cancer, the most crop-intensive counties should have a higher cancer rate than the typical HSA. This difference should be especially clear in a comparison between the most crop-intensive and the least crop-intensive counties.

Table 13 contains a list of the thirty-six counties in Texas with crop-intensity values of 40 percent or greater. By definition these are the counties in which at least 40 percent of the total land area was devoted to harvested cropland in 1949. The counties are ranked within the table from the most crop-intensive county -- Lubbock with about 82 percent -- to the least crop-intensive. Both a simple correlation and a weighted correlation based on respective county population were calculated. Although several sparsely populated counties with ambivalent data could easily outweigh and mask

Table 12. Summary Data of Correlation between Percentage of Total
        Land Area (TLA) Devoted to Harvested Cropland (1949)
        and Average Age-Adjusted Cancer (All Sites) Mortality Rates
        (1969-78) -- HSAs and the State

Geographical Area	Percentage of TLA From Which Crops Were Harvested (1949)	Mortality Rate per 100,000 (1969-78)	Correlation Coefficient
HSA I	31.1	142	-0.14
HSA II	45.6	137	0.33
HSA III	0.7	157	0.62
HSA IV	17.0	140	-0.39
HSA V	29.5	163	-0.10
HSA VI	20.9	148	-0.34
HSA VII	13.6	154	-0.30
HSA VIII	15.1	152	0.22
HSA IX	8.0	162	0.02
HSA X	4.9	165	0.15
HSA XI	13.4	177	-0.38
HSA XII	7.0	152	-0.01
Statewide	16.7	159	-0.13

significant data in one heavily populated county in the simple correlation, the weighted regression should correct for this problem.

Table 14 presents comparable data for the thirty-five Texas counties with less than 2 percent of the total land area devoted to harvested crops. These counties are also listed in ranked format from the virtually crop-free counties such as Winkler (0.0 percent) to Jasper County, with a crop-intensity measure of only 1.9 percent. Simple and weighted correlations were performed.

Table 15 presents the results. There is no apparent association between crop-intensity and total cancer mortality. The four values have a very narrow range from -0.12 to 0.16, which is essentially zero statistically. Moreover, the population of the counties makes no significant difference. Table 15 also notes the overall mortality rates. The most crop-intensive counties had a lower total cancer mortality rate. These results indicate that the areas which probably have greater pesticide exposure may not be at greater risk of total cancer mortality.

Table 13. Ranking of Thirty-Six Texas Counties Most
Crop-Intensive (1949) and Corresponding Average
Age-Adjusted Mortality Rates (1969-78) for All Sites
of Cancer

County	Percentage of Total Land Area From Which Crops Were Harvested	Mortality Rate Per 100,000	Weighting Factor Based on Estimated 1977 Population	Weighted Mortality Rate Per 100,000
Lubbock	82.2	146	200.3	29,243.8
Hale	76.4	129	35.5	4,579.5
Hockley	68.1	141	21.1	2,975.1
Dawson	66.6	132	16.0	2,112.0
Lynn	64.6	127	8.9	1,130.3
Terry	63.4	148	13.9	2,057.2
Cameron	62.7	146	177.1	25,856.6
Lamb	62.0	130	17.7	2,301.0
Floyd	59.8	110	10.5	1,155.0
Swisher	59.1	147	10.2	1,499.4
Parmer	58.6	140	10.2	1.428.0
Castro	57.7	127	10.6	1,346.2
Collin	56.5	155	103.4	16,027.0
Nueces	56.1	174	249.4	43,395.6
Rockwall	56.0	151	10.0	1,510.0
Ochiltree	53.3	135	9.2	1,242.0
Ellis	52.3	147	53.0	7,791.0
Bailey	50.4	114	7.9	900.6
Jones	50.3	137	16.4	2,246.8
Hansford	49.7	136	6.2	843.2
Delta	49.2	138	4.6	634.8
Crosby	48.7	119	8.6	1,023.4
Sherman	47.0	153	3.8	581.4
Willacy	46.6	145	16.8	2,436.0
Hidalgo	46.4	142	235.0	33,370.0
Falls	46.4	140	16.4	2,296.0
Williamson	46.1	146	52.7	7,694.2
McLennan	45.9	149	160.7	23,944.3
Randall	45.8	168	66.6	11,188.8
Hill	45.0	133	22.4	2,979.2
San Patricio	44.8	165	52.6	8.679.0
Runnels	42.3	128	11.4	1,459.2
Fannin	40.5	127	23.1	2,933.7
Carson	40.2	131	10.6	1,388.6
Hall	40.1	150	5.6	840.0
Hunt	40.0	152	49.6	7,539.2

Sources: Texas Department of Health, *Impact of Cancer on Texas, 1980*,
pp. 122-146; and U.S. Bureau of the Census, *Census of
Agriculture (Texas): 1954*, Vol. 1, Part 26, pp. 64-83.

Table 14. Ranking of Thirty-Five Texas Counties Least
Crop-Intensive (1949) and Corresponding Average
Age-Adjusted Mortality Rates (1969-78) for All Sites
of Cancer

County	Percentage of Total Land Area From Which Crops Were Harvested	Mortality Rate Per 100,000	Weighting Factor Based on Estimated 1977 Population	Weighted Mortality Rate
Winkler	0.0	218	9.5	2,071.0
Upton	0.0	124	4.6	570.4
Brewster	0.0	136	7.3	992.8
Crane	0.0	163	4.3	700.9
Kenedy	0.0	20	0.6	12.0
Terrell	0.0	99	1.7	168.3
Culberson	0.1	150	3.5	525.0
Crockett	0.1	165	4.2	693.0
Jeff Davis	0.1	125	1.4	175.0
Loving	0.1	0	0.1	0.0
Val Verde	0.1	172	32.2	5,538.4
Ector	0.2	184	102.7	18,896.8
Reagan	0.2	216	3.5	756.0
Kinney	0.2	133	2.3	305.9
Edwards	0.3	140	2.2	308.0
Presidio	0.3	102	4.7	479.4
Sutton	0.3	175	4.8	840.0
Irion	0.5	102	1.1	112.2
Hardin	0.6	162	37.1	6,010.2
Hudspeth	0.7	125	2.8	350.0
Webb	0.7	149	85.4	12,724.6
Pecos	0.9	132	14.2	1,874.4
Real	1.0	148	2.3	340.4
Sterling	1.1	144	1.0	144.0
Jim Hogg	1.1	124	4.8	595.2
Andrews	1.2	157	11.3	1,774.1
McMullen	1.2	141	0.8	112.8
Aransas	1.3	190	11.1	2,109.0
Tyler	1.4	172	14.6	2.511.2
Montgomery	1.4	195	98.1	19,129.5
Kimble	1.5	141	4.0	564.0
Newton	1.5	105	12.4	1,302.0
Zapata	1.6	119	5.3	630.7
Llano	1.9	161	9.2	1,481.2
Jasper	1.9	146	27.9	4,073.4

Sources: Texas Department of Health, *Impact of Cancer on Texas, 1980*,
pp. 122-146; and U.S. Bureau of the Census, *Census of
Agriculture (Texas): 1954* Vol. 1, Part 26, pp. 64-83.

Table 15. Average Age-Adjusted Mortality rates per 100,000 Population (for All Sites of Cancer) and Simple and Weighted Correlation Coefficients for the Most and Least Crop-Intensive Counties in Texas

Statistic	Thirty-Six Most Crop-Intensive Counties	Thirty-Five Least Crop-Intensive Counties	Statewide (All Counties)
Mortality Rate	150*	167*	159
Simple Correlation Coefficient	-0.12	0.15	-0.13
Weighted Correlation Coefficient	0.16*	0.11*	--

*A weighting factor based on the 1977 county populations as estimated by TDH was used in deriving this figure.

3.3.2.3 Correlation of Farmworker Counties

Rather than using old data on cropped area, it is possible to focus on those counties whose present populations might be more susceptible to the chronic effects of pesticide exposure. As the quotation at the beginning of this report suggests, "it is the farm laborer who experiences most of the occupational disease from pesticides."[107] We therefore assessed the cancer mortality data for those counties with large populations of migrant and seasonal farmworkers. (An individual who has done farmwork within the last five years and has left his or her residence to secure that work is considered to be a migrant farmworker. A seasonal farmworker is defined to be any nonmigrant farmworker who works five months or less per year.) About 43 percent of nonmigrant farmworkers are estimated to work on a seasonal basis[108]

Table 5 above lists the eleven Texas counties which are believed to have the greatest numbers of farmworkers. Cameron and Hidalgo counties in the Lower Rio Grande Valley are estimated to have a very large share of the state's migrant farmworkers. Most of these farmworkers claim residence in the Valley or the Panhandle.

More recent (1982) mortality data were reviewed for these same eleven counties. Five causes of death in addition to cancer also were compared to statewide data. These data are presented in tables 16 and 17. Note that the 1982 mortality data are not age-adjusted (unlike preceding data for 1969-78); one might therefore expect some differences between counties on the basis of age differences in the respective populations.

Table 16. Number of Deaths by Selected Causes in Counties with
Highest Populations of Farmworkers, 1982

County	All Causes	Malignant Neoplasms	Heart Diseases	Pneumonia and Influenza	Chronic Liver Disease	Congenital Anomalies
Bexar (IX)	7,218	1,480	2,455	198	123	74
Cameron (VIII)	1,314	221	359	32	21	11
Deaf Smith (I)	129	25	50	5	0	3
Hale (II)	353	60	119	9	2	2
Hidalgo (VIII)	1,690	284	523	43	22	22
Lubbock (II)	1,441	268	532	34	17	23
Maverick (IX)	153	30	55	2	1	4
Nueces (VIII)	1,937	378	678	32	38	13
Parmer (I)	84	12	28	1	0	3
Starr (VIII)	144	21	60	2	1	3
Webb (VIII)	587	105	194	20	10	3
Texas (Statewide)	111,263	22,470	38,359	2,658	1,329	903

Source: Texas Department of Health, Bureau of Vital Statistics, *Texas Vital Statistics 1982*, pp. 48-61.

Table 17. Mortality Rates (Number of Deaths per 100,000 Population)
         Not Age-Adjusted for Those Counties
         with Highest Populations of Farmworkers, 1982

County	All Causes	Malignant Neoplasms	Heart Diseases	Pneumonia and Influenza	Chronic Liver Disease	Congenital Anomalies
Bexar	703.98	144.35	239.44	19.31 *	12.00 *	7.22 *
Cameron	573.45	96.45	156.67	13.97	9.16 *	4.80
Deaf Smith	582.81	112.95	225.90	22.59 *	0.00	13.55 *
Hale	913.16 *	155.21 *	307.84 *	10.35	5.17	5.17
Hidalgo	541.73	91.04	167.65	13.78	7.05	7.05 *
Lubbock	664.75	123.63	245.42	15.68	7.84	10.61 *
Maverick	429.29	84.18	154.32	5.61	2.81	11.22 *
Nueces	703.45	137.28	246.23	11.62	13.80 *	4.72
Parmer	740.41	105.77	246.80	8.81	0.00	26.44 *
Starr	473.56	69.06	197.32	6.58	3.29	9.87 *
Webb	540.01	96.60	178.47	18.40 *	9.20 *	2.76
Texas (Statewide)	744.51	150.36	256.68	17.79	8.89	6.04

*Asterisk denotes those values exceeding statewide values.

Source: Texas Department of Health, Bureau of Vital Statistics, *Texas Vital Statistics 1982*, pp. 48-61.

Of the eleven counties only one (Hale) has a mortality rate above the statewide rate. This rate difference was quite marked. In addition, only Hale has a mortality rate due to malignant neoplasms and diseases of the heart above the statewide rate. Three counties had excessive pneumonia and influenza mortality rates. Of special interest is the fact that four counties have excessive mortality due to chronic liver disease. Two of these four have very high rates relative to the statewide norm. The liver, which is the primary detoxifying organ in the human body, has been cited as a vulnerable site for chronic disease. Finally, it is worth noting that seven of the eleven counties had high mortality rates as a result of congenital anomalies. Some of these rates are exceptionally high. Parmer's rate of 26 is more than four times that of the state. This could be a result of exposure to teratogenic pesticides or other environmental agents.

Overall, these data also do not suggest a strong correlation between agriculture and cancer. For instance, Hidalgo county has more farmworkers than any other county and is very crop-intensive with a harvested cropland value of 46.4 percent, but both the overall mortality rate and the malignant neoplasm rate are relatively low compared to the rest of the state.

To supplement this analysis, a simple correlation using the 1949 crop-intensity data and 1969-78 total mortality data was run for the eleven "high farmworker population" counties. The correlation coefficient is 0.02. Again, we find no relationship.

3.3.3 Summary of Data Analysis

The morbidity and mortality data reviewed provide virtually no evidence of a positive correlation between pesticide use and chronic (carcinogenic) health outcomes. Simple and population weighted correlations of the 1969-78 mortality data generally indicate the absence of any meaningful correlation between the variables of interest. This was also true for counties with high farmworker populations. A brief review of 1982 mortality rates due to several causes provided some evidence that farming communities may be at higher than normal risk of chronic liver disease and deaths due to congenital anomalies.

In general, our most important finding is that the data are not adequate for the task of assessing the impact of pesticide exposure on cancer mortality rates.   The absence of significant correlations does not mean that pesticide exposure has had no impact on mortality rates. It should be remembered that pesticide usage was not measured directly; cropland intensity served only as a substitute.   For the most part, only total cancer mortality rates were examined. The proxy measures of pesticide use and broad categories of mortality allow too much leeway for narrow (potentially) significant relationships to go undetected within broad nonsignificant relationships. Furthermore, chronic liver disease and congenital anomaly deaths warrant more attention than can be offered herein. This is especially true in light of very recent evidence that the Valley is a "hot spot" for several concerns, including liver damage.[109]

4. PUBLIC POLICIES TO REDUCE HEALTH EFFECTS OF PESTICIDES

4.1 PRESENT POLICIES TO REDUCE EXPOSURE

There are a number of means of reducing exposure to pesticides that are responsive to many of these considerations. Most pesticide labels specify protective clothing that should be worn, and many also indicate a "reentry period," the amount of time after spraying before people should go back into the area. Both these options have disadvantages that make them difficult to use as the only means for reducing pesticide exposure.

4.1.1 Personal Hygiene

With assistance, workers can take some steps to reduce their own exposure to pesticides. NIOSH has advised that washing methods and facilities should be designed to minimize recontamination or exposure. It recommends foot-operated faucets, individual-use towels, dispensable soap, and nonabrasive detergents. The employer needs to take responsibility not only for providing these facilities but for selection of soap, since frequent cleaning increases dermal absorption of pesticides. Creams which counteract this effect should be available.[110]

Furthermore, the value of handwashing may be limited. For example, washing one minute after exposure to parathion is capable of removing one-fourth to one-third of the pesticide. However, after more than one minute, washing is of little benefit. A study of volunteers exposed to parathion found that a soap and water wash for thirty seconds removed only 36-48 percent of the remaining parathion if such cleaning was delayed for six hours.[111] In fact, some evidence suggests that washing four hours after exposure to malathion and eight hours after exposure to parathion may actually increase dermal absorption. Moreover, a complete shower at the end of the working day probably serves no useful purpose in mitigating the impact of exposure to pesticides.[112] However, NIOSH has stated that "showering can be somewhat effective in removing pesticide residues from the body surface."[113] Similarly, washing pesticides from clothing is also very difficult.

The washing facilities described above that TDH has required to be placed in fields and which may soon be required by federal regulation as well were primarily intended to improve worker health through improved sanitation. The data cited here suggest that these facilities, if supplemented by training and provision of appropriate soaps, could also reduce the effects of exposure to pesticides. However, heavy reliance on personal hygiene for pesticide prevention is not warranted by the available data.

4.1.2 Protective Clothing

Protective clothing is specified on most pesticide labels as a means of reducing the effects of exposure. Under federal law, protective clothing means "at least a hat or other suitable head covering, a long sleeved shirt and long legged trousers or a coverall type garment (all of closely woven fabric covering the body, including arms and legs), shoes and socks."[114] Workers are required to be warned orally in their own language and/or by posting at field entry and/or on bulletin boards that they will be working in a field already treated or to be treated and should be told where protective clothing is needed, the time they should leave the field and what to do if exposed. These standards are binding on states, but written in a manner that is very difficult to enforce even at this minimal level. Although states have the right to set and enforce more stringent standards Texas has not yet done so.

However, impermeable protective clothing for farmworkers is not feasible since high temperatures usually require the worker to dress lightly just to dissipate body heat.[115] Furthermore, there is some question about the effectiveness of such clothing. One study found that additional rubberized clothing offered no extra protection against dermal absorption as measured by urinary metabolite testing of exposure. (This test is explained below.)[116] Moreover, pesticides could be detected beneath protective clothing "albeit in small amounts."[117] Another study found penetration of all glove materials used occurred with all pesticides tested within a period of thirty minutes.[118]

The California Department of Health has recommended wearing nonwoven laminar-treated clothing, which is reportedly lightweight, disposable, cool to wear in hot climates, and less permeable to pesticides than tightly woven cotton material. It has also been reported that based on urinary metabolite measurements, workers wearing clothing (shirts, pants, socks, and shoes) dipped in a silicone solution sustain less exposure than workers wearing untreated clothing.[119]

Another disadvantage of protective clothing is that, as noted above, covered skin may absorb more pesticide than uncovered skin. For example, one study found that while the bare palm of the hand absorbed about 11.8 percent of a given amount of applied parathion, the hand absorbed 31.6 percent if "protected" by a canvas glove for a period of eight hours.[120] Other studies have concluded, as commonly believed, that clothing does provide protection. For example, nylon knit gloves used by citrus harvesters in a parathion study allowed only 7 percent of the residue to penetrate the gloves. A similar glove allowed only 6-8 percent penetration of Zolone, a pesticide commonly used on peach crops.[121] It should be emphasized that these figures were apparently derived from "estimates of dermal dose," which were based on a chemical analysis of gauze patches taped to the skin. If the gloves were increasing absorption, the skin itself would have "drained" the patch of available residue, giving these apparently contradictory findings.

4.1.3 Reentry Periods

In general, the regulatory strategy employed in California and nationally to safeguard farmworkers against the hazards of exposure to pesticide residues has been to prohibit entry into treated fields for some prescribed period of time during which residues are presumed to decay to "safe" levels.[122] Reentry periods are established by EPA as part of the registration process and are subject to many of the same scientific uncertainties as other aspects of our understanding of pesticides. Because of these problems, EPA has specified reentry intervals for only twelve active ingredients. However, workers not wearing protective clothing should not be permitted to enter any treated area until sprays have dried or dust has settled. Reentry periods also present a variety of problems in implementation, which are discussed in more detail in the section on policy options in this chapter.

Data on incidents that would confirm or disprove reentry intervals' effectiveness in maintaining worker health are difficult to obtain. Between 1966 and 1979, EPA classified only 86 of 25,500 pesticide incident reports as reentry incidents; of these 39 involved nonfieldworkers. One study of 47 of the fieldworker reentry incidents in nine states (Texas was not one of them) found that reentry problems tended to cluster around particular pesticides. Incidents of exposure to carbamate, aldicarb, and sulfur have occurred within 4 days of application; chlorinated hydrocarbon incidents have occurred with same-day or next-day reentry into treated areas. However, organophosphate incidents have been linked to reentry incidents up to 120 days after application, or well beyond a reasonable reentry period. These delayed reentry incidents have been limited primarily to the use of the insecticides parathion and dialifor in California. The bulk of research has focused on this particular reentry problem.[123]

Nigg and Stamper, citing nine different controlled studies which found very little or no effect of cholinesterase depression or urinary metabolites among workers exposed to dislodgeable residues of organophosphates, concluded that under "normal conditions illness of farmworkers from exposure to residue is aparently not common." Illness results, the researchers have claimed, from (too) early field reentry and/or exposure to unexpectedly high levels of organophosphate oxon metabolites.[124]

Nigg and Stamper posit that the lack of increase in illness is due to the regional character of organophosphate reentry incidents. Four key factors are thought to contribute to a reentry incident five or more days after application: 1. a dusty work environment; 2. use of a sufficiently toxic organophosphate; 3. conversion of the parent compound to its (often more toxic) oxon metabolite, which often occurs under natural conditions; and 4. dry conditions.[125] The presence of the metabolite oxon molecules is the critical feature in all of the cases.[126] These oxon levels vary regionally within the United States. The hot and dusty environment of the centrol California valley makes that region a very troublesome area.[127] To the extent that Texas has the same kind of climate, reentry will pose a problem here as well.[128]

Of particular importance is the point that where particular pesticides exhibit regional "behavior" their relative toxicity to humans may not decrease as time passes following application. Robert Spear has concluded that "the reentry strategy may be inappropriate for environmentally sensitive pesticides because hazard and time are not related in a predictable way. This is an issue of immediate regulatory importance and should be evaluated by the responsible agencies without further delay."[129] In sum, prescription of reentry periods seems to be a commonsense response that should decrease adverse health effects of pesticide exposure, but precise data are not available to form the basis for different reentry periods for various pesticides. In addition, implementation and enforcement of reentry periods may be difficult.

4.2 MONITORING PESTICIDE EXPOSURE

If a pesticide regulation program is intended to protect worker health, then it is necessary to establish a means of monitoring pesticide exposure. This section reviews the scientific and policy bases for monitoring exposure. Just as the causal relationship between pesticide exposure and specific health effects is extremely difficult to prove, so the scientific basis for monitoring has unfortunate gaps. Implementation of a monitoring program would have to take these important gaps into account.

4.2.1 What Is Health Monitoring?

Farmworkers and applicators of pesticides are exposed to substances that are dangerous. Programs that monitor this exposure can help to achieve several of the goals of pesticide programs, including protection of agricultural workers, public health, and the environment, and can assist in enforcing other programs designed to achieve these same goals. If health monitoring, for example, reveals that workers are exposed to high levels of pesticides, this information could be used as the basis of enforcing existing regulations or to show that those regulations are inadequate. At the same time, many of the gaps in the existing information about the health effects of pesticide exposure can be filled by a health monitoring program with good record-keeping facilities. Thus, while monitoring itself is conducted after the fact, it can form the basis for future preventative measures.

There are three kinds of health monitoring activities: recording of acute pesticide poisoning incidents, monitoring exposure levels of workers, and keeping records over long periods of time to discover chronic health effects. Many acute pesticide poisoning incidents are already recorded. Counting them indicates whether the present system is working. The efficacy of this method depends on a variety of factors, including the abilities of doctors to recognize pesticide poisoning (which, as we have already seen, is easily confounded with other diseases); the ease with which health workers can report cases doctors do diagnose; the ability or willingness of agricultural workers to obtain medical care; and the accuracy with which particular pesticides can be named in instances of poisoning. Most of these conditions are not now met; nevertheless, counting programs can provide data on which to base enforcement at the same time that they meet workers' immediate needs for health care.

4.2.2 Health Monitoring through Environmental Testing

Monitoring exposure levels of workers can itself be achieved in two very different ways. One way is simply to measure environmental residues of pesticides; this makes inferences about exposure to people living or working in a residue-ridden area possible. The other is to take samples of blood or urine from workers and test them for evidence of exposure. This is known as biological monitoring. Each form of monitoring has advantages and drawbacks.

Monitoring for residues entails taking samples from affected fields or other areas, running residue tests, and comparing the results to existing standards. Not only are resources required for obtaining samples (although this could be part of routine inspections), but determining which pesticides are present and how much is very expensive and time consuming and not always accurate. Nevertheless, this system is used in several places.

For example, California has a broad monitoring program that includes sampling of air and fields. It conducts:

> Mixer, loader and applicator monitoring to determine potential exposures during pesticide handling and use. Methods include workers' breathing zone samples (respirator exposures), ambient air samples, dermal exposure monitoring with cloth patches, hand washes...

> Pesticide residue studies investigating the degradation of specific pesticides to determine adequacy of reentry tmes for treated fields and other treated areas...residue determinations to assure that workers are not being exposed to harmful residues and that users are complying with established reentry intervals. Methods include collection of foliage, soil, air and/or commodity from treated fields and areas where workers are employed, as well as samples from the skin and clothing of potentially exposed workers.[130]

There are also a few federally recommended or required monitoring programs. For example, the National Institution of Occupational Safety and Health (NIOSH) suggests that air levels of parathion be measured at least monthly. If the levels are high, workers must be notifed, corrective action to lower the levels must be taken, and monitoring must be continued until the levels decrease. The criteria also specify that records of the levels of exposure be kept, possibly for use by physicians. These requirements have been extended to other substances as well.[131] Before DBCP was banned, OSHA required exposure monitoring for it -- except for exposure that resulted "solely from the application and use of DBCP as a pesticide." Thus the rule applied to factory workers, but not to farm workers.[132]

At present, TDA does not monitor the exposure levels of pesticides, or pesticide residues with which farmworkers are likely to come in contact, except in response to a complaint. The difficulties of undertaking a monitoring program and the costs and uncertain results of such a program have deterred the agency so far.

4.2.3 Biological Monitoring

Biological monitoring detects not the pesticides but the evidence of their presence in the body. Cholinesterase monitoring tests a portion of the blood; the other common form of biological monitoring is urine metabolic testing. A 1974 study by the Presidential Task Group on Occupational Exposure to Pesticides concluded that the safety of farmworkers hinged not on the availability and effectiveness of emergency medical care but rather on the preventive measures taken. Prevention requires some form of medical surveillance. The study argued that monitoring the activity of cholinesterase enzymes in plasma and erythrocytes (red blood cells or RBC) is the best available technique for detecting the clinical effects of pesticide exposure, and recommended a program to monitor blood cholinesterase activity levels on a weekly basis.[133]

Since organophosphates act to inhibit cholinesterase, tests designed to measure the levels of cholinesterase activity should reflect exposure to pesticides. There are five basic methods for measuring cholinesterase levels.

One study ranked these five methods preferentially, including the complexity and cost of equipment, the required skill of the analyst, precision, sensitivity, and suitability for routine use. Overall, the colorimetric (DTNB) method was favored, followed by the pH-stat and the Michel methods. The GLC method was least preferred.[134] NIOSH recommends the "Wolfsie and Winter" method -- a modification of the Michel method -- to measure cholinesterase activity in those individuals exposed to organophosphorous insecticides but not carbamate insecticides.[135]

The establishment of individual baseline values of cholinesterase activity during the off season is essential to interpreting cholinesterase test results. Subsequent periodic testing would allow detection of significant exposure whenever an individual's level of cholinesterase decreased appreciably relative to his or her baseline level.[136] However, true baseline values for cholinesterase levels may be difficult to achieve for persons who work with pesticides on a routine basis. For this reason, baseline data should be limited to those workers who have been out of the fields for at least one week. Moreover, several baseline samples should be drawn, with no two samples selected in the same 24-hour period.[137] It has been suggested that farmworkers might be required to carry a card specifying their baseline cholinesterase value; farmers would not be allowed to put a farmworker into the field unless his or her current cholinesterase level was at or near the baseline level.[138]

The conclusions to be drawn from cholinesterase testing are disputed. One reason is that in the course of a normal day of activities, any individual may realize a depression in cholinesterase activity of 15 to 20 percent or more.

[139] Cholinesterase depression in excess of 20 to 30 percent has been regarded by most experts to be of "no physiological consequence."[140] Some laboratory analysts believe that differences of 30 percent or 40 percent would be needed to infer a true effect.[141] On the other hand, one recent EPA manual written for health professionals regards 25 percent depression as "strong evidence of excessive absorption."[142] Systematic depression in cholinesterase levels of whole groups of workers in the same area can be more authoritatively interpreted as indicating excessive exposure.[143]

Furthermore, some analysts believe that the day-to-day exposure to pesticides by farmworkers probably has little, if any, affect on cholinesterase activity.[144] This contention is belied by another study which indicates that farmworkers and their families have a greater tendency to experience relatively higher variations in their cholinesterase activity levels than do nonfarmworkers and their families. For instance, one study found that about three-fourths (75.9 percent) of the "farm" group had experienced red blood cell cholinesterase variation of 20 percent or more compared to about one-half (50.6 percent) of the "nonfarm" group.[145]

Cholinesterase tests of cotton fieldworkers in Arkansas in 1967 and North Carolina in 1971 showed that although no illnesses were reported among the workers monitored, exposure was confirmed. In the Arkansas study, 35 "scouts" were tracked throughout the season. Of these 7 individuals (20 percent) evidenced blood cholinesterase depression of more than 10 percent. However, no decrease was 25 percent or greater. In the North Carolina study, 2 of 28 participants (7 percent) had a reduction in plasma cholinesterase of 25 percent while another 2 (7 percent) had reductions of 40 percent. It was noted that the latter were "careless in habits of cleanliness" and had also entered cotton fields prior to the recommended 48-hour reentry period.[146]

In short, it is difficult to interpret the results of cholinesterase tests because:

1. There is no consensus regarding the amount of depressed cholinesterase activity that is acceptable or unacceptable.

2. There is no consensus regarding the frequency with which cholinesterase testing should be conducted.

3. It is difficult to establish appropriate baseline cholinesterase levels.

4. A depressed cholinesterase level in the field farmworker does not indicate where pesticide exposure occurred, nor does it assist in identifying the .specific chemical involved.

5. Depressed levels may be unrelated to pesticide exposure. Many illnesses, including chronic liver disease, myocardial infarction, hypothyroidism, dermatomyositis, and nutritional disorders can reduce plasma cholinesterase activity. Some disorders such as myasthenia gravis, glaucoma, and emotional problems may be treated with medications which suppress cholinesterase activity.[147] It is also known that genetic factors may contribute to low plasma cholinesterase levels in 3 percent of the population.[148] Finally, cholinesterase can be significantly affected by diet.[149]

In addition to these scientific uncertainties, administration of a cholinesterase monitoring program presents a number of serious implementation difficulties. The main one is that people are usually not anxious to have blood samples taken. This disadvantage is compounded by the fact that the altered level of cholinesterase is not a health problem in itself, just an indication that there may be a problem. Thus individuals have few incentives to submit to testing. Furthermore, there are few if any treatments to be offered if tests show depressed cholinesterase levels; the treatment is to limit exposure until normal healing processes occur. If limited exposure means not working, most people will choose to work. Finally, one of the major advantages of this kind of invasive monitoring is to build up data on which to base future action. There is very little incentive for individuals to undergo unpleasant procedures for the public good. On the other hand, monitoring can warn individuals before they experience serious, perhaps irreversible symptoms.

The procedures recommended by NIOSH for people occupationally exposed to parathion, malathion, methyl parathion, and carbaryl show how difficult program implementation could be. For workers who would be exposed to parathion, NIOSH recommmended that cholinesterase testing be done before employment, again soon after beginning the job to establish a "working baseline," and then routinely. If the levels were abnormal, the employee was to be notified, an industrial hygiene survey was to be done to correct the situation, and, if necessary, the employee was to stop work. This monitoring was to occur in the context of monitoring and recordkeeping of the airborne levels of parathion, and routine medical exams. The employee was to have access to all of this information.[150] The NIOSH document for methyl parathion added to these procedures the findings that the incidence and severity of poisonings increase in hot weather. Thus monitoring should occur more often in the summer.[151]

A related technique for monitoring pesticide exposure is urine metabolite analysis. This technique is superior to cholinesterase monitoring because it is noninvasive, simplifies sample collection, reduces the need for stringent standards of processing, and eliminates much of the guesswork in interpreting the significance of findings.[152] Furthermore, urinary metabolite testing is the most "sensitive and reliable" measure of exposure, yielding quantifiable data when no red blood cell or serum cholinesterase depression could be detected.[153] Like cholinesterase testing, however, urinary metabolite testing also presents logistical, administrative and financial difficulties in monitoring the 2 million agricultural workers in the United States.[154] For example, lab testing must occur within 72 hours of suspected exposure, which presents difficult logistical problems. Since the urinary metabolite test is widely regarded as a more sophisticated and accurate test than cholinesterase testing, research on feasible methods of implementation is clearly warranted.

California has a cholinesterase and urine monitoring program. The three geographical study teams in California which measure pesticide residues and air levels of pesticides also are responsible for biological sampling to quantify known pesticide metabolites in urine and blood samples. However, this testing is generally done in response to a problem or complaint, rather than routinely. Cholinesterase testing thus has been used more as a diagnostic tool than as a monitoring action.

In sum, direct biological monitoring is the only true test of pesticide exposure and can provide important epidemiological evidence of the health effects of such exposure. It suffers from some scientific uncertainties, and from several difficulties in implementation. Among the latter are:

1. The costs of biological monitoring are appreciable; it has not been decided how these costs would be financed.

2. The wide-scale adoption of such monitoring would require many more technicians and laboratories to process specimens numbering in the millions.

3. A system of quality control needs to be devised.

4. Farmworkers would naturally resist the frequent drawing of blood samples.

5. It is unreasonable to expect that farmworkers with depressed levels of cholinesterase would be kept out of the fields during their "recovery period."

6. Even if they could be kept out, doing so would place an added hardship on an already disadvantaged group. (This raises an important question: who would support these farmworkers and their families during this forced layoff?)

7. This monitoring approach does nothing to make the working environment less hazardous. The emphasis is not on making the field safer, but rather on temporarily removing the individual from the environment when the hazard passes some threshold.[155]

The data that can be obtained from a monitoring program are so critical to further regulation, however, that despite these difficulties some kind of monitoring program needs to be undertaken. Data collected can form the basis for new standards. Texas' progress in biological monitoring is discussed in the section below on policy options.

4.3 FEDERAL INITIATIVES IN FARMWORKER HEALTH

4.3.1 PIMS

In an effort to gather a data base on pesticide incidents across the nation, in 1977 EPA established a voluntary Pesticide Incidence Monitoring System (PIMS). EPA has the authority to implement a mandatory system but has not done so.[156] Labs involved with pesticide use, poison centers, crisis centers, medical centers, and university facilities were designated to gather reports for the federal agency. Information on incidents came from poison centers, doctors, hotlines for the public, and various agencies. Records on all calls about possible pesticide problems were forwarded to Washington and used to identify persistent problems associated with specific products, and to determine where and how major hazards were occurring in order to prevent future incidents.[157]

The PIMS system has been criticized both for not reflecting the true extent of pesticide-related health problems and for exaggerating those problems. Some reports document underreporting,[158] and others believe PIMS data tends to overreport incidents.[159] The Texas Pesticide Hazard Assessment Project (PHAP) summary in 1980 included this disclaimer:

> These reports do not represent a uniform sample of the five states in which the Texas PHAP attempts to gather information. Neither are they a proper sample of all the circumstances in which pesticides might be involved, nor are they a sample of all the pesticides . . . The Texas reports best reflect events in the eastern and extreme southern parts of the state...[160]

In any region the number of PIMS reported incidents will almost surely greatly underestimate the actual numbers of problems associated with pesticides. Many if not most cases of mild to moderate exposure will not be reported to any authority. As noted, symptoms of poisoning closely mimic reactions to other common diseases, or may not occur until a period of time after exposure and thus not be connected by the person to the pesticidal source. For the same reason, even when symptoms are described to a physician, poisoning may not be diagnosed. Also, many incidents are settled between applicators (especially in homes) and those harmed, without formal complaints. Given the voluntary nature of PIMS and the low public awareness of poisoning symptoms, there is little incentive to document pesticide health problems, or to investigate where such problems may be suspected. As a result, the incidents that are eventually recorded at a PIMS center can be well below actual occurrences in the community.[161]

Migrant worker health clinics have frequent contact with pesticide-related health complaints, yet only one nonfatal incident of poisoning was documented by a migrant clinic nationwide in a year of PIMS data (1979). Underreporting is clearly illustrated by a detailed study of farmworkers in South Florida. During July 1978 and June 1979, four nonfatal incidents involving a total of twenty-four people were reported to PIMS from Palm Beach County, Florida. Researchers who investigated more thoroughly discovered at least forty-two documented cases of pesticide abuse and poisoning in that county during the same time.[162] These cases were not reported through the PIMS process.

Since PIMS centers accept and record all calls whether substantiated or not, they may be documenting incidents that would otherwise be dismissed for insufficient evidence. In the 1980 Texas PHAP summary, of the 2,335 total reports only 153 (6.5 percent) were confirmed as cases of a specific chemical leading to illness, death, or damage to property. In the other incidents, although potential hazards existed, no illness or damage was documented. One estimate is that only 1 of every 10 or 15 Hotline inquiries is a documented incident.[163] The collected PIMS calls were not easily sorted for authenticity and were open, according to some, to misinterpretation.

Overreporting also occurred because there was no clear guide as to how to define an "incident." If one application of a pesticide resulted in ten independent calls, in some centers that became one incident with ten people harmed while in another record it was ten separate incidents. With no set procedure, results from different centers could not be reliably compared.

PIMS figures have been used in regulatory actions by EPA. When the new

Administration took over in 1981, state agencies' and pest control associations' complaints about the lack of validity of PIMS data led to severe cuts in the program's budget. Collection centers had to shut down operations, close hotlines, and/or reduce the amount of monitoring that could be maintained. However, in the new 1984 budget EPA has refunded a revised PIMS, called PIMS II.[164] Currently the proposals for the new program are undergoing peer review.

Major changes have been proposed to satisfy complaints about the past unreliable figures. In the new system as proposed, all reported incidents will be given an authentication number (AN) according to generally accepted criteria. If an incident is reported but neither the active ingredient nor the extent of injury is known and there is no corroboration by either lab tests or medical diagnosis, it will get an authentication number and code of AN/0. Succeeding grades will designate incidents from AN/1 to AN/4, based on whether potential hazards are known, and/or whether the active ingredient is proven to cause the stated symptoms, and/or whether medical or lab diagnosis has been obtained, A label will represent a defined level of confirmation for each complaint.[165]

Another proposed change would create a network of agencies and medical personnel by having the major enforcement agency for pesticide use in each state collect the data and be given the first opportunity to apply an authentication number before it is sent to pesticide monitoring centers. The state agency would be given a small amount of EPA funding to pay for coordinating the data. The purpose of involving these agencies is to ensure that they are alerted to statewide health problems associated with pesticides and are able to monitor all incidents, whether filed through formal complaints or documented by health personnel. It has been argued that putting more responsibility on state agencies will bring the reporting system closer to where the usage takes place; currently PIMS is federally operated and removed from the source of actual problem solving.[166] It would be the task of PHAPS to be the prime contact for the agency and through them to connect the various poison centers, medical personnel, labs, and hospital emergency rooms to improve the flow of incident reporting to PIMS. The system currently being discussed would still be voluntary; therefore, increased educational outreach will be necessary to stimulate cooperation and consistent recordkeeping.

After the state agency applies a code the PHAP could review the data and possibly revise the codes if necessary, then send the record to the new computerized central information bank at Mississippi State University. There it may have a third review by an independent auditor before being entered into the national PIMS file. There are quality assurance control audits recommended via sampling techniques throughout the system.[167]

Plans were to begin to set up the PIMS II in March 1984. Seven PHAPS have been proposed to run a trial program for one year based in Hawaii, California, Texas, Colorado, Iowa, Mississippi, and Florida. It is anticipated that in subsequent years at least one or two more areas could be set up. with each center responsible for a five- or six-state region.

Any PIMS-type system will still monitor only for acute pesticide poisonings. However, figures on these extreme health problems associated with pesticide use would help state agencies assess where and what the immediate hazards are in Texas. While research on methods to diagnose and monitor long-term problems must continue, it is evident that lack of knowledge of current health incidents has been a major stumbling block to interpreting and applying epidemiological and laboratory studies. A more accurate and comprehensive PIMS reporting system will increase public awareness of pesticide health effects, especially in the medical community, and is an approach to gaining a better understanding of health conditions in the state.

4.3.2 Other Federal Initiatives

As already noted, federally sponsored migrant worker clinics in Texas have recently begun cholinesterase testing on routine basis. This is part of a "new urgency" by the U.S. Department of Health and Human Services (DHHS) "to address the human health implications of our widespread reliance on chemical pesticides." In a 1981 memorandum to the Region VI Migrant Health Program coordinator, the Office of Migrant Health (within DHHS) emphasized that "all Migrant Health Grantees [centers] are required by law to develop the capacity to detect, treat and report adverse health effects resulting from exposure to pesticides." Funds were allocated to Region VI (Texas, Louisiana, New Mexico, Arkansas, and Oklahoma) to develop and implement a training program geared toward assisting all migrant health projects in the development of pesticide health hazard management programs. The ultimate goal of such a program "is to provide all clinics serving migrant and seasonal farmworkers, applicators, and anyone else working in close proximity to pesticides with the ability to detect and document a base-line cholinesterase enzyme level in an individual's medical record. This base-line level can then be used for comparison purposes in the event of a possible pesticide exposure.[168]

Training was conducted in 1982 and 1983 through the Regional Cluster

Training Center (RCTC), located in Albuquerque, New Mexico. The training

workshops were geared toward all clinic personnel -- administrators and

outreach workers as well as physicians, nurses, and lab technicians. Medical

and lab personnel were taught diagnostic methods that can be administered with

the varied toxicity levels and nonspecific symptomatology which can be

manifested with pesticide poisoning. Administrative, outreach, and support

staff were introduced to poisoning symptom identification, protective

equipment, and techniques in developing effective community education

programs. Aside from the actual training program, the Office for Migrant

Health also published a document entitled *A Guide to the Development of a*

Pesticide Health Hazard Management Program. This document is intended for the

use of any ambulatory (non-hospital) health care provider, to promote

awareness of pesticide-related health problems within the medical

community.[169]

As part of its charge to monitor the effects of pesticide use on humans and

the environment, EPA established a special unit within the Office of

Pesticides and Toxic Substances called the Pesticide Farm Safety Staff. The

unit was created during the Carter Administration and reports directly to the

Director for Pesticide Programs. Broadly, the staff office was established to

provide a focus for agricultural pesticide safety issues. Within EPA, the

Farm Safety Staff performs coordination and oversight functions to assure that

farm safety issues are adequately addressed (including development of policy

and program recommendations), and also serves as an agency ombudsman for farm

safety concerns. As part of the federal-state partnership for pesticide

enforcement, the Farm Safety Staff promotes intergovernmental coordination of

activities (including federal, state, and other) designed to provide for safe

93

use.[170]

4.4 NEW TDA POLICIES

Until December 8, 1983, there was no program or office within the TDA specifically focused on the needs of farmworkers in this agricultural state. Even the TDA complaint form reflected the invisible status of workers by not including health problems as one of the reasons for reporting a complaint. In December 1983 the first Farmworker Safety Pesticide Task Force was established. Headed by Ed Gutierrez, newly appointed co-ordinator of the Farmworker Program, the task force is an internal group whose purpose is to gather information throughout the state on ways to improve farmworker conditions. Members of the task force include representatives of the enforcement area of TDA's Agricultural and Environmental Sciences Division (Ken Kadlec), the assistant deputy for Regulatory and Consumer Services (Ron White), legal counsel (Jim Butler and Delores Alvarez Hibbs), and the supervisor of field operations (Lamarcus Johnson).

Initially four areas were under consideration for immediate action: worker protection standards, farmworker education, applicator training, and modification of labeling on pesticides. The group sponsored informal hearings around the state in order to solicit opinions from a variety of sources on how TDA can modify or develop farmworker regulations. Suggestions were made by people involved with production and sales of agricultural chemicals, growers, aerial and ground applicators, dealers of pesticides, farmworker representatives such as the Texas Pesticide Research and Education Project, United Farm Workers, Texas Rural Legal Aid, and Catholic charities groups, environmental activists such as the Sierra Club and also suburban activist groups formed in response to concerns about pesticide drift into nonfarm areas. A report incorporating the testimony developed by the Farmworker Safety Pesticide Task Force will then be submitted to the agriculture commissioner for approval. If it is approved, TDA will publish proposed regulations in the Texas Register for notice and comment, and through the administrative process new regulations will be adopted, probably in autumn 1984.[171]

In addition to developing new standards, TDA has hired a toxicologist to undertake more basic research on potential hazards associated with the use of pesticides in Texas. This will be part of a larger effort to analyze farmworker conditions.[172]

Public education is a priority for the Farmworker Program. Pamphlets in both Spanish and English, videotapes for public clinics and farmworker organizations, and public service announcements for rural TV and radio are some of the ways TDA will be alerting workers and their families to potential hazards. A statewide hotline for pesticide calls was proposed which could help disseminate information, document incidences of health problems, and notify TDA investigators of possible problems that need attention.

Regulations proposed during the spring 1984 program hearings may eventually include new reentry intervals and field posting requirements and increased training for applicators (both licensed and certified applicators as well as persons working with pesticides without licensing or certification such as mixers, loaders, and field scouts). Equipment and inspection specification may be modified to reduce worker exposure risks. Health care workers could be given training in agricultural regions to deal with pesticide health effects, a particularly important issue if workman's compensation laws are enacted by the Texas legislature. Monitoring such as cholinesterase or urinary metabolite testing may be proposed for workers in contact with pesticides. Another important agency action could be requiring collection of use records of all pesticides from applicators or dealers. Currently there is no way to determine how much pesticide or which specific chemicals are being applied. Such information is necessary in documenting worker health complaints, since individual workers often will not know what pesticides were applied in their work sites.

The problem of drift has caused suburban groups to get involved in pesticide issues. Buffer zones may be needed to protect schools and hospitals and homes located near agricultural fields, but details on how to establish and enforce such zones remain to be developed.

TDA's Farmworker Safety Pesticide Task Force reflects a new commitment to farmworkers in the state. The agency's attention to the needs of those who work in the fields should help protect the health of all. The next section discusses several policy approaches that will enable state agencies to carry out this new commitment.

4.5 POLICY OPTIONS TO REDUCE HEALTH EFFECTS OF EXPOSURE

TDA has a variety of options in formulating its new policies to limit adverse effects of pesticides. Options that include limiting the numbers of potential applicators, enforcing existing regulations more stringently, and other such measures are discussed in the companion volume. Here we consider a few options to implement general policies that are most directly concerned with worker health: reentry periods and collection of data about exposure.

4.5.1 Regulating Farmworkers' Exposure: Reentry

The risk of exposure to pesticide poisoning is greatest at the time of application. For this reason it is generally agreed that farmworkers and others should not be in or immediately adjacent to these fields during the time of application. There is disagreement and some lack of knowledge as to when farmworkers should be permitted or obliged to "re-enter" the fields following application.

Three policy options for reentry are: 1. continue the present (EPA based) system; 2. monitor residue levels prior to permitting reentry; and 3. adopt reentry intervals based on specific toxicity, residue behavior, and exposure mechanisms as they exist in agricultural areas within Texas.

The present regulatory strategy is to prohibit entry into treated fields for some prescribed period of time during which residues are presumed to decay to "safe" levels. However, there are guidelines for only a few chemicals, and the data on which these are based are old. An environmental monitoring system, such as the one used in California, actually measures existing residues. Again, the absence of underlying scientific data on residue levels, especially ones that are acceptable in the specific climate of Texas, may reduce the utility of this kind of program.

The flexible reentry option assumes that reentry regulations should vary as the risk of exposure varies. This regulatory strategy also requires a sound scientific understanding of the variables that affect reentry times, including crop, pesticide(s), weather, and other human and environmental factors. TDA's toxicologist proposes to undertake the necessary research. He estimates that with a small staff and a modest budget most of the Texas-specific reentry data base could be realized in about two or three years.

A flexible reentry strategy recognizes the changing character of pesticide residues. Some areas and some pesticide-crop settings in Texas at some times of the year may require very stringent reentry intervals (as is the case in California). Special reentry regulations may be needed to the extent that agricultural areas in Texas have the following conditions:

1. very dry periods -- especially around harvest time;

2. extended dry periods and a large quantity of particulate matter below 50 micrometers;

3. high use of organophosphates -- especially parathion and dialifor;

4. special considerations such as crop pest management practices, work practices, and labor relations.

Once a general reentry strategy is adopted, several specific issues in implementation arise. Two of the most important of these are the way in which time intervals are defined and posting.

It may be more practical and enforceable to specify reentry intervals in terms of "working days" rather than "24- or 48-hour periods." Requiring a minimum of one full working day between application and reentry might eliminate substantial regulatory guesswork in investigating any complaints of violations. For instance, a field treated late in the afternoon on Tuesday could not be entered until Thursday morning. Most importantly, the full working day buffer ensures that the 24-hour period cannot be compressed.

Posting supplements reentry regulations by informing workers of areas affected by reentry requirements. One possibility is to require "posting notices" to be aired over the radio or published in the newspaper to inform local residents what substances are scheduled to be applied in specific areas. This unusual procedure is similar to one used to announce public hearings, but would probably have a more limited audience, especially among workers, and would require a detailed knowledge of local land ownership to be effective.

A more common procedure is to require signs or flags in fields. Growers may object to signs and flags because of their cost, because their liability for damages when signs are stolen or deteriorated is unclear, and because there is no clear evidence that posting is effective. Advocates of posting say that workers have a right to know the risks to which they are exposed and argue that without posting reentry regulations are ineffective.

One decision that affects the cost of posting is the specificity of the notices. Pesticide-specific signs would require that a more extensive inventory of signs be kept on hand. A more reasonable and manageable approach might be a color or code number to signify only the class of pesticide used (e.g., organophosphate). This, of course, would require an effort to educate farmworkers about the significance of the color or number. But any posting strategy would require some corresponding educational program.

Another decision that affects costs is the placement of notices. Costs can also be expected to be higher the closer together signs are required to be placed. It seems reasonable that all fields should have signs posted at well-defined entry points (e.g., gates, if fenced) and corners. Perimeter posting at some fixed distance may serve no added useful purpose with very large, open fields that have multiple entry points if some lesser number of signs strategically placed at access roads and junctions provides adequate notice.

4.5.2 Usage Data

Texans do not have reliable basic information on pesticides in the state. Texas has no program like that of California for assessing how much of any particular chemical is used even on a county-by-county basis.[173] As we have noted, there is also no consistent reporting of health problems caused by pesticidal chemicals. Effective regulation depends on accurate knowledge of the locations where pesticides are being applied and on dependable monitoring of the health of people working with these compounds.

The California Departments of Food and of Agriculture have a pesticide use report with which a farm operator must document each pesticide application, its dosage and volume, the crop treated and location, and the date and time of application for every pesticide that has a reentry safety interval assigned by the State.[174] This record-keeping system can then be used when investigating reported misuse or overexposure. Texas now requires dealers to document and save for two years records of their sales of restricted pesticides.[175] Dealers do not routinely have to submit these figures to TDA, although TDA does have the right of access. Texas law also requires licensed applicators to maintain records of pesticide use, with the amount of detail at the discretion of the agency.[176] Again, a copy must be sent to TDA if requested. Without going beyond its existing authority, TDA could require these forms to be sent in. Coded and entered on the computer, the data could form the basis for improved regulation.

4.5.3 Medical Data

Usage data are not enough for a complete assessment of problems caused by pesticides. The State also needs adequate systems of reporting health and safety complaints by workers affected by the chemicals. As described above, existing voluntary systems for obtaining data are unreliable. California has overcome this problem in part through a system that allows an employed worker to go to any physician and get examined for any illness or injury the worker believes is work-related. The physician must submit the work injury form in order to get reimbursed for the examination, regardless of the diagnosis. Texas might adopt a similar system which provides an incentive to health care providers that would encourage better documentation of incidents, or TDA could make pesticide health complaint reports mandatory. The advantage of the incentive method is that enforcement is built into the system, whereas a mandatory system entails significant enforcement costs.

Applicators and/or growers could also be held legally responsible for providing health care to workers exposed to pesticides, again alerting the medical community to incidents of pesticide related health complaints. Each employer would make arrangements with a physician to treat any cases of poisoning experienced by workers. This requirement would put a financial burden on the employers, yet it would provide added incentive to them to assure a safe work environment for employees who use toxic substances. There will be more incentive to obtain accurate reports of acute pesticide poisoning if workman's compensation is extended to agricultural workers.

Physicians and other health care personnel would need to be trained in the proper diagnosis and treatment of pesticide complaints; the protocol for such training already exists in numerous programs run for migrant clinics. The Department of Agriculture, the Department of Health, or the State Board of Medical Examiners could distribute information to county health associations at minimal cost (relative to the benefits) for dissemination throughout the state. A strong commitment from each of these organizations would be necessary in order to ensure adequate training and interest on the part of individual health care personnel throughout the state. One approach could include pesticide-related health training in continuing education requirements for physicians.

TDA has considered setting up a state hotline which would be widely publicized so that even individuals not likely to get medical attention could obtain some aid. In addition to giving information about treatment, the hotline personnel could document reported symptoms and which chemicals were involved, although accurate information would be difficult to obtain over the telephone. Such a system is relatively inexpensive to operate, but requires wide publicity to succeed.

All these plans would improve reporting of acute pesticide poisonings, but would not monitor chronic illnesses. Biological monitoring is almost the only way to achieve that goal.

4.5.4 Biological Monitoring

We have discussed the ways in which biological monitoring can serve to protect worker health directly by discouraging further exposure and to provide a set of data for monitoring exposure over long periods. In assessing monitoring options, it is important to remember that it is a *post hoc* method that can detect pesticide exposure only after it has occurred. Its role in prevention is limited to telling exposed workers not to be further exposed. It could play a role in enforcement as well, with consistently lowered cholinesterase levels indicating that existing regulations are not being followed or are inadequate.

The testing options can be listed on a continuum of frequency:

1. Require those tests as diagnostic measures whenever a person reports symptoms of pesticide poisoning, or when an accident happens which may have resulted in pesticide exposure.

2. Require routine biological monitoring -- cholinesterase testing or urinary metabolite testing -- for all workers who use or might reasonably be expected to come into contact with pesticides that cause the biological changes that these monitoring tools detect.

3. Design a plan of comprehensive medical surveillance for all those who use or come into contact with pesticides in their work. Such a plan would include, but not be limited to, routine cholinesterase or urinary metabolite testing. It would also include regular general physical examinations, and more specific testing for possible chronic effects of pesticides (cytogenetic testing, for example).

Results of these kinds of programs could be used in several ways to:

1. Report adverse findings to TDA. This reporting would have limited use for prosecution of violations, but could be used to obtain postmarketing information for registration reviews.

2. Tie enforcement actions to the results of tests. If routine monitoring of workers' cholinesterase levels at a particular site indicated unacceptable exposure to pesticides, then TDA would conduct more frequent and thorough inspections at that place and/or impose sanctions. Or, if TDA found a violation, and at the same time workers were showing changes in cholinesterase levels, then TDA would move the violation into a more serious category.

3. Require farmworkers to carry a card specifying their baseline cholinesterase value; farmers would not be allowed to put a farmworker into the field unless his or her current cholinesterase level was at or near the baseline level.[177]

4. Require the employer to provide a genuine alternative to working at the same hazardous job for workers found to have unacceptable cholinesterase levels. One option would be to require the employer to provide an opportunity to transfer to another available and comparable job (not offering something the worker can't do or wouldn't want to do) at the same rate of pay. Another would be to guarantee workers' compensation. (This generally only covers lost work time of a week or more, however.)

5. Provide a mechanism to ensure that the worker will not face or fear retaliation from the employer for submitting to these tests or seeking medical care.

There are a variety of problems with implementing any of these programs of biological monitoring. We have noted that people are usually not anxious to have blood or urine samples taken, especially since the altered level of cholinesterase is not a health problem in itself. We also noted other disadvantages of participating in a program, including the likelihood that the only treatment is avoiding exposure, which entails a loss of income. The major incentive to participate in a program is that workers will learn when they are seriously at risk. With so many disincentives, however, a program might have to be mandatory to succeed. This creates a terrible administrative burden and puts the onus on the worker to comply.

Employers similarly have no incentives to provide monitoring or encourage employees to participate in programs unless monitoring is linked to enforcement. Even if fines and sanctions could be issued on the basis of testing, the fact that workers move from field to field would make assignment of responsibility to a particular employer extremely difficult. The possibility that sanctions could be invoked would actually give employers incentives to discourage workers from submitting to tests.

Cholinesterase testing on a routine basis has recently been initiated in federally sponsored migrant worker health clinics in Texas.[178] A test is performed on any patient who exhibits pesticide poisoning symptoms, and on anyone who has worked in a field up to seven days prior to the clinic visit.

The testing program is still in the initial stage of development. Clinic personnel were trained during 1982 and 1983, and clinics began testing as equipment was received during 1983. By November 1983, only eleven of the sixteen clinics in the state had begun testing. The number of tests per clinic ranges from 4 to 432. (The wide range is due to the differing start-up dates of each clinic's program.) The total number of tests performed by November 1983 was 952. Of these, only 4 showed abnormal levels, with 1 patient requiring hospitalization. It should be emphasized that the data are acknowledged to be questionable, since in 1983 clinic personnel were still relatively unfamiliar with testing equipment and testing procedures in general.[179] The program has also been criticized for underparticipation. Because farmworkers are reluctant to undergo testing, many do not "confess" that they have been in the field within seven days.[180]

The lab in San Benito, Texas will soon begin a urinary metabolite screening system at the migrant worker clinic in Hidalgo (the Hidalgo County Health Corporation). Since the urine metabolite test is more sophisticated than the cholinesterase test, it will be used as a check on the validity of the cholinesterase testing program. The Hidalgo County clinic was chosen because of its proximity to the San Benito lab, since the urine metabolite test must be done within 72 hours of suspected exposure.

4.6 CONCLUSION

This report had two major purposes: first, to provide a succinct review of the existing literature on health effects of fieldworkers' exposures to pesticides; and second, to review existing policies and consider possible policy options. The review of the literature confirmed that there are many, many unsettled scientific questions that inhibit our understanding of the relationships between pesticide exposure and health. In addition to limits on our basic knowledge of the mechanisms by which pesticides might affect human health, there are serious limits on our abilities to infer causal relationships because of the absence of good aggregate data on amounts of pesticides used, numbers of workers, and incidence of diseases.

As we have noted, however, taken together all the anecdotal, epidemiological, and laboratory studies seem to suggest that there are some serious health effects from pesticide exposure. Even in the absence of perfect data on these effects, workers and many others believe that policies should be formulated that will protect them to the greatest possible extent. Our review of the policies presently in force suggests that they afford rather limited protection to workers. Formulation of new policies that will be sure to be effective, however, depends on increased understanding of the relationships between exposure and health. Therefore, we have placed a great deal of emphasis on policy options that call for better data collection. Collection and analysis of data will take many years, however; in the interim, commonsense policies such as field sanitation and enforcement of reentry periods can help to minimize exposure. Applicator and worker education, an option discussed in the companion volume, must supplement other policies, since enforcement of pesticide regulations is so difficult in the field.

Individuals who come in contact with pesticides must have enough information to take some responsibility for their health into their own hands.   Ensuring that workers have appropriate information is a major task, one that TDA is beginning to undertake. Health workers, farmers, and others must assist if such an effort is to succeed. We hope that this volume itself helps increase people's understanding of this difficult subject.

NOTES

[1]Irma West, "Occupational Experiences with Pesticides in California" (paper presented at International Short Course on the Occupational Aspects of Pesticides, Center for Continuing Education, University of Oklahoma, Norman, November 20-22, 1963), p. 11.

[2]U.S. General Accounting Office, *Stronger Enforcement Needed against the Misuse of Pesticides*, CED-82-5, October 15, 1981, p. 2.

[3]U.S. Department of Agriculture, Economic Research Service, "Pesticides: Weighing Benefits against Risks," *Farmline* 1, no. 9 (December 1980): 16.

[4]U.S. Department of Agriculture, Economic Research Service, *Agricultural Outlook*, ao-92, "IPM: The Route to Efficient Pest Control," by Katherine Reichenfelder (Washington, D.C., October 1983), p. 22.

[5]Ibid.

[6]U.S. Environmental Protection Agency, *National Household Pesticide Usage Study, 1976-1977* (Washington, D.C., November 1979), p. 76.

[7]Ibid.

[8]Study reported in U.S. General Accounting Office, *Stronger Enforcement Needed*, p. 72.

[9]Karen Mountain and Dr. Mary Walker, *Pesticides in Texas: The Facts* (Austin: Texas Rural Health Field Services Program, University of Texas at

Austin School of Nursing, 1981).

[10]U.S. General Accounting Office, *Stronger Enforcement Needed*, p. 72.

[11]Dallas Morning News, *Texas Almanac 1982-83* (Dallas: A. H. Belo Corp., 1983), p. 517.

[12]Telephone interview by C. Miller with Dr. Rodney Holloway, Entomologist, Texas A&M University, College Station, Texas, November 3 and 4, 1983.

[13]Portions of this section are adapted from U.S. Environmental Protection Agency, Office of Pesticide Programs, *Pesticide Protection: A Training Manual for Health Personnel*, by J. Davies, March 1977.

[14]William T. Keeton, *Biological Science* (New York: W. W. Norton and Co., 1980), pp. 423-29.

[15]U.S. Environmental Protection Agency, "Recognition and Management of Pesticides Poisonings," Technical Report EPA 540/9/80/005, January 1982; National Institute for Occupational Safety and Health,"Occupational Exposure during the Manufacture and Formulation of Pesticides," July 1978, pp. 9-10.

[16]"Miosis" means extreme smallness of the eye's pupil. "Encephalitis" means inflammation of the brain.

[17]Ibid.

[18]Florida Rural Legal Aid Services, Inc., *Danger in the Field* (May 1980).

[19]National Institute for Occupational Safety and Health (NIOSH), *Criteria for a Recommended Standard: Occupational Exposure during the Manufacture and Formulation of Pesticides* (Washington, D.C.: U.S. Government Printing Office, July 1978), pp. 132-35.

[20]Interview with Rebecca Harrington, Director, United Farm Workers, AFL-CIO, Texas Chapter, Austin, November 29, 1983. She reports that, when asked on a weekly basis, fieldworkers almost invariably offer no complaints about exposure to pesticides. Harrington attributes much of this silence to the workers' general view that a skin rash or headache is nothing to get upset about.

[21]Ibid.; and Ephraim Kahn, "Pesticide-Related Illness in California Farm Workers," *Journal of Occupational Medicine* 18, no. 10 (October 1976): 693-96.

[22]Jay Feldman, "Statement of Jay Feldman, National Coordinator, National Coalition Against the Misuse of Pesticides, before the Subcommittee of Department Operations, Research and Foreign Agriculture, Committee on Agriculture, U.S. House of Representatives, October 6, 1983." Mimeograph.

[23]Ephraim Kahn, "Pesticide-Related Illness," p. 694.

[24]U.S. Environmental Protection Agency, Office of Pesticide Programs, *National Study of Hospital-Admitted Pesticide Poisonings* (Washington, D.C.: U.S. Government Printing Office, 1976); also U.S. EPA, Office of Pesticide

Programs, *National Study of Hospitalized Pesticide Poisonings, 1974-1976*, by Eldon Savage et al., Epidemiologic Pesticides Studies Center, Colorado State University (Washington, D.C.: U.S. Government Printing Office, 1980).

[25]S. T. Caldwell, and M. T. Watson, "Hospital Survey of Acute Pesticide Poisoning in South Carolina, 1971-1973," *Journal of the South Carolina Medical Association* 71, no. 8 (August 1975): 249-52.

[26]Interview with Philip Zbylot, MD, Chief, Preventive and Community Medicine, Austin-Travis County Health Department, Austin, October 13, 1983.

[27]NIOSH, *Recommended Standard*, p. 245.

[28]Ibid., p. 115.

[29]Ibid., pp. 138-81.

[30]Leon F. Burmeister, "Cancer Mortality in Iowa Farmers, 1971-1978," *Journal of the National Cancer Institute* 66, no. 3 (March 1981): 461-64. The cancers studied affected the stomach, prostate, and lip, and included leukemia, non-Hodgkins lymphoma, and multiple myeloma.

[31]Leon F. Burmeister et al., "Selected Cancer Mortality and Farm Practices in Iowa," *American Journal of Epidemiology* 118, no. 1 (July 1983): 72-77.

[32]Ibid., p. 73.

[33]Mabuchi, MD, et al., "Cancer and Occupational Exposure to Arsenic: A Study of Pesticide Workers," *Preventive Medicine* 9, no. 1 (January 1980): 51-77.

[34]E. Barthel, "Increased Risk of Lung Cancer in Pesticide-Exposed Male Agricultural Workers," *Journal of Toxicology and Environmental Health* 8 (November-December 1981): 1027-40.

[35]California Department of Health Services, *Literature Review on the Toxicological Aspects of DBCP and an Epidemiological Comparison of Patterns of DBCP Drinking Water Contamination with Mortality Rates from Selected Cancers in Fresno, California, 1970-1979: A Report to the California Department of Food and Agriculture* (Berkeley: California Department of Health Services, June 1, 1982), pp. 2-53.

[36]Harrison Stubbs, John Harris, and Robert C. Spear, *Mortality of California Agricultural Workers: 1978-79*, Report No. NCOHC 83-B-3 (Northern California Occupational Health Center, June 1983.)

[37]Donald P. Morgan, Lawrence I. Lin, and Heidi H. Saikaly, "Morbidity and Mortality in Workers Occupationally Exposed to Pesticides," *Archives of Environmental Contamination and Toxicology* 9, no. 3 (1980): 349-82; Helen Wang and Brian MacMahon, "Mortality of Pesticide Applicators," *Journal of Occupational Medicine* 21, no. 11 (November 1979): 741-44.

[38]NIOSH, *Recommended Standard*, pp. 182-89.

[39]David Schwartz, MD, et al., "Parental Occupation and Birth Outcome in an

Agricultural Community." Mimeograph, 1980.

[40]Richard Louv, "High Birth Defect Rate Found among Farmworkers' Children," *Sacramento Union*, November 12, 1980.

[41]Telephone interview with Dr. Molly Coye, pediatric researcher, Santa Barbara, California, March 14, 1984.

[42]NIOSH, *Recommended Standard*, pp. 183-89.

[43]Clements Association, Inc., *Chemical Hazards to Human Reproduction* (Washington, D.C.: Council on Environmental Quality, 1981), pp. IV-22, IV-27, A-28.

[44]Whorton, M. D., et al., "Testicular Function in DBCP Exposed Pesticide Workers," *Journal of Occupational Medicine* 21, no. 3 (March 1979): 161-166.

[45]NIOSH, *Recommended Standard*, pp. 189-95.

[46]S. Lings, "Pesticide Lung: A Pilot Investigation of Fruit-growers and Farmers during the Spraying Season," *British Journal of Industrial Medicine* 39 (1982): 370-76.

[47]Florida Rural Legal Aid Services, Inc., *Danger in the Field*, pp. 36, 51, 52.

[48]Morgan et al., "Morbidity and Mortality," p. 349.

[49]Keith T. Maddy, "Farm Safety Research Needs in the Use of Agricultural Chemicals" (California Department of Food and Agiculture, Draft Document), p. 3.

[50]Federal Working Group on Pest Management, *Occupational Exposure to Pesticides: Report to the Federal Working Group on Pest Management from the Task Group on Occupational Exposure to Pesticides* (Washington, D.C.: U.S. Government Printing Office, January 1974), p. 29.

[51]Ibid., pp. 32, 34.

[52]Philip L. Martin, "Labor-Intensive Agriculture," *Scientific American* 249 no. 2 (October 1983): 54-59.

[53]William J. Popendorf and John T. Leffingwell, "Regulating OP Pesticide Residues for Farmworker Protection," *Residue Reviews* 82 (1982): 125-201.

[54]NIOSH, *Recommended Standard*, p. 55.

[55]Howard I. Maibach and Robert Feldman, "Systemic Absorption of Pesticides through the Skin of Man," *Occupational Exposure to Pesticides*, pp. 120-27.

[56]NIOSH, *Recommended Standard*, p. 56.

[57]William J. Popendorf and Robert C. Spear, "Preliminary Survey of Factors Affecting the Exposure of Harvesters to Pesticide Residue," *American Industrial Hygiene Association Journal* (June 1974): 374-80.

[58]Tracy Freedman and David Weir, "Polluting the Most Vulnerable," *Nation*, May 14, 1983, p. 602.

[59]Telephone interview with William J. Popendorf, Industrial Hygienist, Institute of Agricultural Medicine, University of Iowa, Iowa City, Iowa, February 6, 1984.

[60]Robert C. Spear, "Farmworker Exposure to Pesticide Residues: Reflections on Differential Risk," *Banbury Report 11* (1982): 67-76.

[61]Ibid., pp. 72-73.

[62]Freedman and Weir, "Polluting," pp. 602-3.

[63]Carol K. Redmond, "Sensitive Population Subsets in Relation to Low Doses," *Environmental Health Perspectives* 42 (December 1981): 137-40.

[64]Ecologic studies attempt to relate group differences in exposure to group differences in frequency of adverse effects (e.g., a comparison of rates of cancer among agricultural workers with rates among nonagricultural workers).

[65]Marvin A. Schneiderman, *Environmental Health Perspectives* 42 (December 1981): 33-38.

[66]Ibid.

[67]There is an additional problem in the area of pesticide-related health

research. Based on journal article listings, a substantial amount of research in the field is published in Soviet Union and Soviet-bloc journals. Unfortunately, in practical terms, much of this information is unavailable. Article abstracts have sometimes been translated into other languages, including English; the published information, however, is often limited and difficult to evaluate in the absence of detailed methodology.

[68]U.S. Department of Agriculture, Crop Reporting Board, *Crop Production -- August 1983* (August 1983), p. B-24.

[69]Texas Crop and Livestock Reporting Service, *Texas Agricultural Facts*, September 2, 1983.

[70]Governor's Office of Migrant Affairs, *Migrant and Seasonal Farmworker Population Survey: Final Report*, Report No. 76-01 (Austin, July 15, 1976), pp. 20-21.

[71]Ibid., pp. 3, 27-29.

[72]Texas Rural Legal Aid, Inc., and United Farm Workers, AFL-CIO, Farm Worker Advocacy Project, *Farm Workers and Workers' Compensation in Texas* (Austin, August 1983), p. 4.

[73]Governor's Office of Migrant Affairs, *Population Survey*, pp. 43-44.

[74]Ibid.

[75]Martin, "Labor-Intensive Agriculture," pp. 56-57.

[76]Texas Rural Legal Aid, Inc., and United Farm Workers, *Farm Workers*, p. 4.

[77]Texas Department of Health, Occupational Health Program, *Environmental Standards for Sanitation at Temporary Places of Employment, Pursuant to Texas Sanitation and Health Protection Law*, June 1983, secs. 289.95.a.5, 289.97.a.4, 289.97.a.5, and 289.98.c.

[78]Interview by Gary Watts with Troy W. Lowry, Registered Sanitarian, Migrant Labor Housing Sanitation Branch, Texas Department of Health, Austin, February 17, 1984.

[79]Jess F. Kraus et al., "Epidemiologic Study of Physiological Effects in Usual and Volunteer Citrus Workers from Organophosphate Pesticide Residues at Reentry," *Journal of Toxiocolgy and Environmental Health* 8 (1981): 169-84, as cited in Texas Rural Legal Aid, Inc., *Texas State Plan for Farmworkers* (February 1982), p. 47.

[80]Texas Rural Legal Aid, Inc., *Texas State Plan*, p. 47.

[81]Ibid., p. 44.

[82]Texas Department of Health, *The Texas State Health Plan* (Austin, 1982), p. 496.

[83]U.S. Department of Health and Human Services, *Health -- U.S. 1980*, Report

No.   (PHS)81-1232 (Washington, D.C.: December 1980), p. 79, as cited in Texas Department of Health, *Health Plan*, p. 501.

[84]Texas Department of Health, *Health Plan*, p. 501.

[85]G. A. Reich et al., "Characteristics of Pesticide Poisoning in South Texas," *Texas Medicine* 64 (September 1968): 56-58.

[86]D. A. Smith and J. S. Wiseman, "Epidemiology of Pesticide Poisoning in the Lower Rio Grande Valley in 1969," *Texas Medicine* 67 (February 1971): 56-58.

[87]Ibid.

[88]Ibid., p. 56.

[89]U.S. Environmental Protection Agency, 1976, and U.S. Environmental Protection Agency, 1980.

[90]Telephone interview by C. Miller with Mike Ellis, Director, Southeast Texas Poison Center, Galveston, Texas, November 3, 1983.

[91]Texas Tech University Health Services Center, *1980 Pesticide Incident Summary*, Pesticide Hazard Assessment Project, San Benito, Texas., Unpublished data.

[92]National Rural Health Council, *Pesticide Use and Misuse: Farmworkers and Small Farmers Speak on the Problem* (Washington, D.C.: Rural America, 1980),

p. vii.

[93]Karen Mountain and Mary Walker, *Pesticides in Texas: The People and the Issues* (Austin: Texas Rural Health Field Services Program, 1981).

[94]Federal Working Group on Pest Management, *Occupational Exposure to Pesticides*, pp. 46-47.

[95]Robin Alexander, Policy Research Project Class Presentation, Lyndon B. Johnson School of Public Affairs, University of Texas at Austin, September 20, 1983.

[96]National Rural Health Council, *Pesticide Use and Misuse*.

[97]U.S. EPA, *Diagnosis and Treatment of Poisoning by Pesticides* (Washington, D.C.: n.d.), p. 6.

[98]Interview with Phillip Zbylot, Austin, October 13, 1983.

[99]Telephone interview with Ephraim Kahn, Epidemiologist (retired), California Department of Health Services, Berkeley, November 18, 1983.

[100]Telephone interview with Robert C. Spear, Professor of Environmental Health, University of California at Berkeley, November 30, 1983; and with Herbert Nigg, Associate Professor, University of Florida, November 17, 1983.

[101]Telephone interview with Herbert Nigg, November 17, 1983.

[102]Texas Rural Legal Aid, Inc., *Texas State Plan*, p. 154.

[103]Larry C. Clark at al., "Cancer Mortality and Agricultural Activity: An Association with Cotton Production and Large Farms," in Paul E. Leaverton (ed.) *Environmental Epidemiology* (New York: Praeger Publishers, 1982), p. 5.

[104]Eleanor J. Macdonald and Evelyn Heinze, *Epidemiology of Cancer in Texas: Incidence Analyzed by Type, Ethnic Group, and Geographic Location* (New York: Raven Press, 1978).

[105]Interview with Richard A. Beauchamp, Environmental Epidemiologist, Texas Department of Health, Austin, May 14, 1984.

[106]Ibid.

[107]West, "Occupational Experiences."

[108]Governor's Office of Migrant Affairs, *Migrant and Seasonal Farmworkers in Texas* (Final Report). Austin, Texas: GOMA, July 1976, pp. 20-29

[109]Telephone interview with Jim Dobbins, University of Texas Medical Branch at Galveston, May 14, 1984.

[110]NIOSH, *Recommended Standard*, p. 273.

[111]Ibid., p. 57.

[112]Federal Working Group on Pest Management, *Occupational Exposure to Pesticides*, p. 70.

[113]NIOSH, *Recommended Standard*, p. 274.

[114]40 CFR 170, 39 FR 16890, May 10, 1974.

[115]Federal Working Group on Pest Management, *Occupational Exposure to Pesticides*, p. 69.

[116]C. A. Franklin, et al., "Correlation of Urinary Pesticide Metabolite Excretion with Estimated Dermal Contact in the Course of Occupational Exposure to Guthion," *Journal of Toxicology and Environmental Health* 7 (1981): 715-31.

[117]Ibid., p. 730.

[118]NIOSH, *Recommended Standard*, p. 255.

[119]Ibid.

[120]Federal Working Group on Pest Management, *Occupational Exposure to Pesticides*, p. 70.

[121]William J. Popendorf et al., "Harvester Exposure to Zolone (R) (Phosalone) Residues in Peach Orchards," *Journal of Occupational Medicine* 21, no. 3 (March 1979): 189-94.

[122]Robert Spear, "Technical Problems in Determining Safe Reentry Intervals," *Journal of Environmental Pathology and Toxicology* 4 (1980): 293-304.

[123]Herbert N. Nigg and James H. Stamper, "Regional Considerations in Worker Reentry," *American Chemical Symposium Series: Pesticide Residues and Exposure* no. 182 (1982): 62-63.

[124]Ibid., pp. 63-65.

[125]Ibid., p. 67.

[126]Herbert N. Nigg, J. A. Henry, and James H. Stamper, "Regional Behavior of Pesticide Residues in the United States," *Residue Reviews* 85 (1983): 257-76.

[127]Nigg and Stamper, "Regional Considerations," p. 68.

[128]A study of parathion decay in California orange groves by Popendorf and Leffingwell found a strong correlation between residue oxidation (conversion of parathion to paraoxon) and dry, stable weather conditions, supporting the "regional hypothesis" (William J. Popendorf and John T. Leffingwell, "Natural Variations in the Decay and Oxidation of Parathion Foliar Residues," *Journal of Agricultural Food Chemicals* 26, no. 2 [1978]: 437-41).

[129]Spear, "Technical Problems," p. 301.

[130]California Department of Occupational Safety and Health, "California

Worker Health and Safety Program Statement Outline, 1982-1983," Sacramento, September 1982.

[131]See, for example, NIOSH 76-206, "Criteria for a Recommended Standard . . . Occupational Exposure to Malathion," June 1976.

[132]OSHA Health and Safety Standards, General Industry Standards 2 CFR SS 1910.1044 (1983).

[133]Federal Working Group on Pest Management, *Occupational Exposure to Pesticides*, pp. 34, 71.

[134]Federal Working Group on Pest Management, *Occupational Exposure to Pesticides*, p. 139.

[135]NIOSH, *Recommended Standard*, p. 371.

[136]Federal Working Group on Pest Management, *Occupational Exposure to Pesticides*, p. 71.

[137]Ibid., p. 109.

[138]Ibid., p. 71.

[139]Ibid., p. 49.

[140]Ibid., p. 107.

123

[141]Janice Yager et al., "Components of Variability in Blood Cholinesterase Assay Results," *Journal of Occupational Medicine* 18, no. 4 (April 1976): 242-44.

[142]U.S. EPA, *Recognition and Management*, p. 4.

[143]"Medical Supervision of Pesticide Workers, Guidelines for Physicians" (State of California, Department of Heath Epidemiological Stuides Laboratory, 1974).

[144]Telephone interview with Robert C. Spear, November 30, 1983.

[145]Federal Working Group on Pest Management, *Occupational Exposure to Pesticides*, p. 50.

[146]Ibid., p. 34.

[147]Ibid., p. 110.

[148]U.S. EPA, *Pesticide Poisonings*, p. 4.

[149]Federal Working Group on Pest Management, *Occupational Exposure to Pesticides*, p. 110.

[150]"Criteria for a Recommended Standard . . . Occupational Exposure to Parathion" (NIOSH 76-190, June 1976). Similar recommendations were made in "Criteria for a Recommeded Standard . . . Occupational Exposure to Malathion"

(NIOSH 76-205, June 1976).

[151]"Criteria for a Recommended Standard . . . Occupational Exposure to Methyl Parathion (NIOSH 77-106, April 1976).

[152]Kraus et al., "Epidemiologic Study," p. 183.

[153]Franklin et al., "Correlation," p. 727.

[154]Popendorf and Leffingwell, "Regulating OP Pesticide Residues," p. 128.

[155]Federal Working Group on Pest Management, *Occupational Exposure to Pesticides*, pp. 71-72.

[156]Florida Rural Legal Aid Services, Inc., *Danger in the Field*.

[157]Telephone interview by C. Miller with Vaughn Cox, Lab Director, Texas Poison Hotline, Texas Tech Lab, San Benito, Texas, October 28 and November 3, 1983.

[158]Florida Rural Legal Aid Services, Inc., *Danger in the Field*.

[159]Telephone interview by C. Miller with Harold Trammel, Director, National Pesticide Telecommunications Network, South Carolina, Noverber 15 and December 5, 1983.

[160]Texas Tech University Health Services Center, *1980 Pesticide Incident*

Summary.

[161]Telephone interview by C. Miller with Dr. Tony Mullhagen, Director, Texas Pesticide Hazard Assessment Program, San Benito, Texas, November 21 and December 5, 1983.

[162]Florida Rural Legal Aid Services, Inc., *Danger in the Field.*

[163]Telephone interview by C. Miller with Dr. James Minyard, Project Manager, PIMS II, Mississippi State University, Mississippi State, Mississippi, November 28 and December 5, 1983.

[164]Telephone interview by C. Miller with Willa Garner, Assessment Division, EPA, Washington, D.C., November 22, 1983.

[165]Minyard interview, November 28 and December 5, 1983.

[166]Trammel interview, November 15 and December 5, 1983.

[167]Minyard interview, November 28 and December 5, 1983.

[168]Rita Pino and Vincent Gomez, "Health Hazard Management Training Program," *Texas Rural Health Journal* (May-June 1983): 27-29.

[169]U.S. Department of Health and Human Services, Public Health Service, Health Services Administration, Bureau of Community Health Services, Office for Migrant Health, *A Guide to the Development of a Pesticide Health Hazard*

Management Program (Washington, D.C.: Bureau of Community Health Services, March 1982).

[170]U.S. Environmental Protection Agency, Office of Pesticide Programs, Pesticide Farm Safety Staff, *Policy Statement and Action Plan*, March 1983.

[171]Interview by C. Miller with Ed Gutierrez, Coordinator of Farmworker Program at TDA, February 16, 1984.

[172]Telephone interview by C. Miller with Vaughn Cox, TDA Toxicologist, February 17, 1984.

[173]California Administrative Code, Title 3, Chapter 4, Subchapter 1, Group 2, #2481 *Records*.

[174]California Administrative Code, Title 3, Chapter 4, Subchapter 1, Group 2, #2481 *Records*.

[175]Texas Pesticide Laws, Subchapter D, sec. 76.075 *Records*.

[176]Chapter 76. Pesticide Law, Subchapter E, sec. 76.114.

[177]Federal Working Group on Pest Management, *Occupational Exposure to Pesticides*, p. 71.

[178]Migrant health clinics are located in rural areas throughout the United States and provide care to migrants and seasonal farmworkers and their

dependents. Centers usually provide a comprehensive range of care, from primary medical services to supplemental care such as dental and vision services. The centers are coordinated through the Migrant Health Program in the Bureau of Health Care Delivery and Assistance, within the Public Health Service of the U.S. Department of Health and Human Services.

[179]Telephone interview by A. Rade with Salvador Mier, Regional Program Consultant, Migrant Health Program, Bureau of Health Care Delivery and Services, Public Health Service, U.S. Department of Health and Human Services, October 31, 1983.

[180]Interview by A. Rader with Rebecca Harrington, Director of United Farm Workers, AFL-CIO, in Texas, November 29, 1983.